The Urban Book Series

The Urban Book Series is a resource for urban studies and geography research worldwide. It provides a unique and innovative resource for the latest developments in the field, nurturing a comprehensive and encompassing publication venue for urban studies, urban geography, planning and regional development.

The series publishes peer-reviewed volumes related to urbanization, sustainability, urban environments, sustainable urbanism, governance, globalization, urban and sustainable development, spatial and area studies, urban management, transport systems, urban infrastructure, urban dynamics, green cities and urban landscapes. It also invites research which documents urbanization processes and urban dynamics on a national, regional and local level, welcoming case studies, as well as comparative and applied research.

The series will appeal to urbanists, geographers, planners, engineers, architects, policy makers, and to all of those interested in a wide-ranging overview of contemporary urban studies and innovations in the field. It accepts monographs, edited volumes and textbooks.

Indexed by Scopus.

Genet Alem Gebregiorgis · Stefan Greiving ·
Ally Hassan Namangaya · Wilbard Jackson Kombe
Editors

Planning Cities in Africa

Current Issues and Future Prospects of Urban Governance and Planning

Springer

Editors
Genet Alem Gebregiorgis
International Planning Studies
TU Dortmund University
Dortmund, Germany

Stefan Greiving
Institute for Spatial Planning
TU Dortmund University
Dortmund, Germany

Ally Hassan Namangaya
School of Spatial Planning and Social Sciences
Ardhi University
Dar es Salaam, Tanzania

Wilbard Jackson Kombe
Institute of Human Settlement Studies
Ardhi University
Dar es Salaam, Tanzania

ISSN 2365-757X ISSN 2365-7588 (electronic)
The Urban Book Series
ISBN 978-3-031-06549-1 ISBN 978-3-031-06550-7 (eBook)
https://doi.org/10.1007/978-3-031-06550-7

This Springer imprint is published by the registered company Springer Nature Switzerland AG
The registered company address is: Gewerbestrasse 11, 6330 Cham, Switzerland

Foreword

In his essay collection entitled L'Intention poétique (Poetic Intention), the Martinican writer Édouard Glissant (2010) wrote:

> It was in Ibadan, Nigeria, that I discovered and experienced what is referred to as the force of a people. That city … extends over an infinity of little houses or cabins whose agglomeration finds its meaning in the full of the night. Then the light of street stalls, the music, the tumult, and the very odors populate and install it in a familiar electricity. City of alleys and shanties, of dim lights and smoke…. A city without monuments but where perfection crackles in the presence of the world.

When I first read this, I immediately understood and felt this familiar electricity that comes from the force of a people in African cities. African urban studies have long faced challenges of lumping disparate narratives of urbanisation across the continent to make a broader case for Africa's cities' significance in global urban studies. Nevertheless, at the risk of being accused of this lumping, there are themes in common, including this one: that there is a familiar electricity, a vibrancy, an energy, that comes from the way ordinary people live their lives and create space in the region's cities (Simone, 2004).

Urban planners in Africa often do not know what to do with this familiar electricity, because it accompanies other commonalities. One of these is the circumstance of urbanisation without growth. Even if the macroeconomic situation for most African countries has been improving, for cities in the region the rate of population growth still outstrips the rate of economic growth in most cases. Secondly, while the region's cities rank low on scholarly indexes of globalisation, the creep of global influences is palpable to any urban observer, in the shopping malls, foreign-owned luxury apartment towers and regional headquarters for international organisations or transnational firms that abound (Myers, 2020). At the same time, large numbers of people in these very same cities have been struggling to survive, to meet even the most basic needs for food, shelter, and water, well before the COVID-19 pandemic began. And even beyond these dynamics, urban planners in Africa continue to face numerous other daunting challenges.

As the editors of this book note in their introduction, these challenges include the ever-expanding impacts of global climate change, the struggle for affordable housing, the provision of urban services, informality, the persistence of slums and poverty, and many more urgent problems. The chapters in this volume are organised around three of these many challenges, with an eye towards the practice of the planning profession and the training of professional planners—how to localise planning theories and models to build more collaborative and participatory approaches, how to change the state of planning education and build capacity, and how planning fits within a multi-level governance approach to the key problem areas of urban life. The most important contribution that this book makes to the literature is in its fostering of an "understanding of cities and urbanisation in Africa from their own sociocultural and historical perspectives", from Morocco to South Africa, from Ghana to Ethiopia, with material on cities in several other countries as well. It is, perhaps, the key challenge for planners on the continent to learn how to work with the "force of a people" from these endogenous and indigenous perspectives, as many of the chapters palpably demonstrate.

The volume's origins lie in analysing African planners' efforts to address the urgent arena for urban policy presented by global climate change. Addaney and Cobbinah (2019, p. 4) noted the truism that "Africa's contribution to global climate change is comparably negligible" yet "it remains the most affected region". It is also under-researched in urban terms: their book was the first comprehensive survey for the region's cities. Addaney (2019, p. 482) noted that in Africa, the urban vulnerabilities to climate change "are well-documented" but "less attention has been paid to how the city government plans to adapt to climate change and enhance the resilience of the local population". This volume adds important case studies to the burgeoning literature that is finally paying attention to this realm of planning on the continent.

Throughout all three thematic areas of the book, one can follow the struggle over inclusivity in planning—in terms of including local people and perspectives in theory building and modelling, including collaborative and participatory approaches in planning school curricula, and practising inclusivity in the governance and implementation of planning schemes. Planning across Sub-Saharan Africa often has had a poor record of inclusivity in practice from colonial times onwards (Watson, 2009 and 2014; Silva, 2020). African cities have several decades of experience with the pervasive uncertainty about the roles of the public, private and popular sectors in governance, governing, and politics which underpin the struggle over inclusivity. Currently, the sharpness of meaning and clarity of triumphal vision around the Global North's intertwined ideas of urban democracy and urban neoliberalism are evidently in doubt, from Kandahar to Kyiv to Khartoum. Militaries and corporate elites share tactics to crush popular interests, in many countries, surely not in Africa alone, often leaving well-honed participatory urban planning schemes in tatters

Yet, at the same time, substantive social movements for radical democratisation from the grassroots have also taken hold in urban areas, even in some of the same countries experiencing revanchist authoritarianism, such as Sudan. One must not "romanticise" the "global protest wave" of the twenty-first century. But sometimes, from the grassroots up, African cities are sharing paths towards "inventing new visions of democracy and development in which popular interests come first" in addressing urban crises (Branch and Mampilly, 2015, p. 6). The chapters of this volume will prompt further discussion of the ways in which a grassroots politics in urban Africa can transform how planners work to strengthen the hands of responsive, developmentalist, and collaborative leaders in governments and urban communities—to imagine where planning can tap into the "familiar electricity" with inclusive deployment of that "force of a people".

Wethersfield, USA Garth Myers

References

Branch A. and Mampilly Z (2015) Africa uprising: Popular protest and political change. Zed Books, London

Cobbinah PB and Addaney M eds (2019) The geography of climate change adaptation in Urban Africa. Springer International Publishing, Cham

Glissant É (2010 [1969]) Poetic intention (trans: N. Stephens) Nightboat Books, Callicoon NY

Myers G (2020) Rethinking urbanism: Lessons from postcolonialism and the global south. Bristol University Press, Bristol, UK

Silva CN (2020) Ancient, Colonial, and Post-colonial Urban Planning in Africa: An introduction. In: Silva CN (ed.), Routledge Handbook of Urban Planning in Africa, Abingdon and New York. Routledge, pp 1–11

Simone A (2004) People as infrastructure: Intersecting fragments in Johannesburg. Public Culture, 16(3): 407–429

Watson V (2009) Seeing from the south: Refocusing urban planning on the globe's central urban issues. Urban studies, 46(11), pp 2259–2275

Watson V (2014) African Urban Fantasies: Dreams or Nightmares?. Environment and urbanization, 26(1): 215–231

Acknowledgements

With this book, we aim to contribute to the discourse on planning theory, education, and urban governance-related issues in Africa. Above all, we aim to draw the attention of young African scholars and call upon them to take part in the discourse and contribute not only to the perspective from the south, but also to localise the discourse to the varied and dynamic cities of the continent. In this, we were assisted through the support and collaboration of several people and institutions. The chapters were developed from selected papers submitted to the 5th AAPS (Association of African Planning Schools) Conference. This would not have been possible without the generous financial, institutional, and individual support and encouragement that we enjoyed. Therefore, we would like to acknowledge and express our deep gratitude to all those who directly or indirectly contributed to the production and publication of this book.

We would like to express our gratitude to our colleagues at IRPUD—Institute of Spatial Planning (TU Dortmund University), special thanks go to Dr. Joahnnes Lückenkötter, coordinator of the capacity building module within the GOPLAREA (Governance and Planning of Resilient Cities in East Africa) research project, for his input, coordination of panel discussions and his unwavering support for the creation of this book. We extend our appreciation to our student assistants, Viola Ricksgers and Miriam Schertl. We are, indeed, touched by, and admired their patience and disposition throughout their time with the research project GOPLAREA. We are also grateful to and thank the student assistants at IRPUD, Stefan Andreas Adamson, Maria Matschulski and Jannik Josef Sebastian May, for offering their time and support during the conference.

The process to produce this book was technically complex, as the funding was linked to the GOPLAREA project which was finalised last year in December. But the reviewing and editing of the chapters took much longer time than planned. In this situation, the language editor Ms. May Koot went out of her way to assist us. We are most grateful and would like to thank her for her careful editing, time and above all for extending her service beyond the schedule agreed in the work contract.

We are grateful to the institutes and colleagues who joined the organisation and scientific committees of the 5th AAPS Conference. In particular, we would like to acknowledge and express our appreciation to the research group IPS (International Planning Studies) at our university, and its head Prof. Sophie Schramm, for the generous organisational support and reviewing conference abstracts. We are grateful for the approval of AAPS to host the association's conference and the support from the individual members of its String Committee. Special thanks go to Prof. Nancy Odendaal, University of Cape Town, South Africa, and Prof. Daniel Inkoom, Kwame Nkrumah University of Science and Technology, Ghana.

We are greatly indebted to the colleagues who supported us in reviewing conference papers and moderation of panels and sessions of the conference. We are particularly grateful for the support of Dr. Raffael Beier, International Planning Studies at TU Dortmund University; Steven Bland, Climate Change Planning Unit, Urban Planning and Design Branch, UN-Habitat; Francis Dakyaga, Dombo University of Business and Integrated Development Studies, Wa, Ghana; Dr. Juan Du, IRPUD, TU Dortmund University, Germany; Dr. Lia Gabremariam Woldetsadik, EiABC—Ethiopian Institute for Architecture, Building and Construction, Addis Ababa University, Ethiopia; Dr. Priscila Izar, School of Architecture and Planning, Centre for Urbanism and Built Environment Studies, University of Witwatersrand, South Africa; Prof. Alphonce Kyessi, Institute of Human Settlement Studies (IHSS), Ardhi University, Dar es Salaam, Tanzania; Dr. Dawah Magembe-Mushi and Dr. Tatu Mtwangi-Limbumba, Centre for Continuing Education and Institute of Human Settlement Studies (IHSS), respectively, Ardhi University, Dar es Salaam, Tanzania; Dr. George Okechukwu Onatu, University of Johannesburg, South Africa; Dr. Samwel Alananga Sanga, Department of Business Studies School of Earth Sciences, Real Estate, Business and Informatics (SERBI), Ardhi University, Dar es Salaam, Tanzania; Dr. Wolfgang Scholz, IRPUD, TU Dortmund University, Germany; Dr. Julia Wesely, University College, London, UK; and Dr. Nestory E. Yamungu, Department of Geography, College of Social Sciences, University of Dar es Salaam, Tanzania.

We are grateful to all the authors and keynote speakers who took their time to prepare papers and presentations for the conference. We are indeed honoured to have had the speeches of Prof. Robert Kiunsi, School of Environmental Sciences at Ardhi University, Dar es Salaam, Tanzania; Prof. Garth Myers, Centre for Urban and Global Studies, Trinity College, Hartford, USA; Dr. Wilma Nchito, Department of Geography, University of Zambia, Lusaka, Zambia; Prof. Amira Osman, Faculty of Engineering and the Built Environment at the Tshwane University of Technology, Pretoria, South Africa; and Prof. Michael Poku-Boansi, Department of Planning at Kwame Nkrumah University of Science and Technology, Kumasi, Ghana.

Last but not least, we wish to thank BMBF (German Federal Ministry for Education and Research) for funding the research project GOPLAREA, including the conference and publication of this book.

Contents

Editors and Contributors

About the Editors

Dr. Genet Alem Gebregiorgis is a Lecturer in the Department of International Planning Studies of the Faculty of Spatial Planning at TU Dortmund University. She has worked as a lecturer at Dohuk University in Iraq and at the Ethiopian Civil Service University in the Urban Management Masters' Programme. Genet Alem also has several years of experience practising as a professional architect and planner. In Ethiopia, she has worked for the Federal Urban Planning Institute, Addis Ababa Works and Urban Development Bureau and Office for the revision of the Addis Ababa Master Plan. Alem holds a Ph.D. in Spatial Planning from TU Dortmund, an M.Sc. in Urban and Regional Planning from TU Dortmund and Dar es Salaam University and a degree in architecture from The Technical University of Havana José Antonio Echeverría (La CUJAE) in Havana, Cuba. Genet Alem is member of Network-Association of European Researchers on Urbanization in the South (N-earus), Planners for Climate Action (P4CA) and TRIALOG e.V.—Association for Scientific Research into Planning and Building in the Developing World.

Prof. Dr. Stefan Greiving has been Head of the Institute of Spatial Planning at TU Dortmund University since 2011. He holds a degree in spatial planning and a Ph.D. in urban planning. His postdoc thesis (habilitation) focused on spatial planning and risk. Prof. Greiving has been PI of about 40 international and 80 national projects on disaster risk management, adaptation to climate change, and regional development. He is the author of about 250 publications. Prof. Greiving is a member of the German Academy for Territorial Development and a member of the chamber of architects.

Prof. Ally Hassan Namangaya has been in practice as a land use planner since he completed his training. He is an Associate Professor at Ardhi University. He obtained his Ph.D. at TU Dortmund University in 2010. He is currently Dean in the School of Spatial Planning and Social Sciences. He has undertaken research and been published in peer-reviewed journals on issues such as urban competitiveness and land-based financing; rural land use conflicts; community-based natural resource management; the application of GIS in land use planning; and the linking of rural communities to tourism value chains. He has extensive experience as a consultant and in research on the reforms of urban planning practices and the incorporation of digital technologies into urban planning. He has led the preparation of master plans for Kibaha, Geita and Morogoro, and several neighbourhood and institutional master plans (estate development plans). In rural development planning, he has worked on community engagement and land use planning for out-grower development project entities such as RMI, Eid environment, Eco-Energy and Southern Agricultural Growth Corridor of Tanzania (SAGCOT). He is the non-executive founding Director of the PAMS Foundation, which is dealing with anti-poaching and conservation in the Selous Niassa Wildlife Corridor and the Ngorongoro Conservation Area and its surroundings.

Wilbard Jackson Kombe is a Professor at the Institute of Human Settlement Studies at Ardhi University in Tanzania. Kombe received his Ph.D. from the Technical University of Dortmund. Since his graduation, he has extensively collaborated with various institutions, both locally and internationally, in research and training activities. His recent research work has focused on land management and urban governance, informal urbanisation in poverty, urbanisation and public policy analysis. He has been extensively involved in applied and basic research activities in numerous countries in Sub-Saharan Africa. This includes research activities undertaken in Tanzania, South Africa, Zimbabwe, Kenya, Uganda, Mozambique, Zambia, Namibia, Botswana and Ethiopia. He has also led a number of multi-disciplinary research teams and projects undertaken in Tanzania and other countries in the region.

Contributors

Achamyeleh Gashu Adam Institute of Land Administration, Bahir Dar University, Bahir Dar, Ethiopia

Samwel S. Alananga School of Earth Sciences, Real Estate Business and Informatics, Ardhi University, Dar es Salaam, Tanzania

Genet Alem Gebregiorgis International Planning Studies (IPS), TU Dortmund University, Dortmund, Germany

Geetika Anand Indian Institute for Human Settlements, Bengaluru, India

Milousa António New University of Lisbon, Lisbon, Portugal

Raffael Beier TU Dortmund University, Dortmund, Germany

Wafae Bouallala Institut National d'Aménagement et d'Urbanisme, Rabat, Morocco

Innocent Chirisa Department of Urban & Regional Planning, University of the Free State, Bloemfontein, South Africa;
Department of Demography Settlement & Development, University of Zimbabwe, Harare, Zimbabwe

Francis Dakyaga Faculty of Spatial Planning, Technische Universität Dortmund, Dortmund, Germany;
Department of Local Governance and City Management, SD Dombo University of Business and Integrated Development Studies, Wa, Ghana;
Department of Urban and Regional Planning (URP), Ardhi University, Dar es Salaam, Tanzania

Behailu Melesse Digafe Institute of Land Administration, Bahir Dar University, Bahir Dar, Ethiopia

Nandini Dutta Indian Institute for Human Settlements, Bengaluru, India

Stefan Greiving Institute for Spatial Planning (IRPUD), TU Dortmund University, Dortmund, Germany

Jennilee M. Kohima Department of Architecture and Spatial Planning, Namibia University of Science and Technology, Windhoek, Namibia

Wilbard Jackson Kombe Institute of Human Settlement Studies, Ardhi University, Dar es Salaam, Tanzania

Dawah Magembe-Mushi Ardhi University, Dar es Salaam, Dar es Salaam, Tanzania

Abraham R. Matamanda Department of Geography, University of the Free State, Bloemfontein, South Africa

Ramadhani Matingas Ardhi University, Dar es Salaam, Dar es Salaam, Tanzania

Mengiste Abate Meshesha Institute of Land Administration, Bahir Dar University, Bahir Dar, Ethiopia

Ally Hassan Namangaya School of Spatial Planning and Social Sciences, Ardhi University (ARU), Dar es Salaam, Tanzania

Verna Nel Department of Urban & Regional Planning, University of the Free State, Bloemfontein, South Africa

Gebeyehu Belay Shibeshi Institute of Land Administration, Bahir Dar University, Bahir Dar, Ethiopia

Lia Gabremariam Woldetsadik Ethiopian Institute of Architecture, Building Construction and City Development, Addis Ababa, Ethiopia

List of Figures

List of Tables

Chapter 1
Planning Cities in Africa—Current Issues and Future Prospects of Urban Governance and Planning: An Introduction

Genet Alem Gebregiorgis, Ally Hassan Namangaya, Stefan Greiving, and Wilbard Jackson Kombe

Abstract African urbanisation is at the crossroad. Despite the ever-growing urban challenges and rapid transformation of cities in Africa, there is a positive trend of knowledge production and dynamic policy reforms aiming at a better management of urbanisation and related development fields. The discourse on current African urban challenges and prospects is calling for a change of perspective in understanding urban Africa from its own sociocultural and historical context. Scholars, for instance, (Connell, Plan Theory 13:210–223, 2014), (Robinson, J. (2006) Ordinary cities: between modernity and development. London; New York: Routledge (Questioning cities); Robinson, Int J Urban Reg Res 35:1–23, 2011)) and (Watson, V. (2009) 'Seeing from the South: Refocusing Urban Planning on the Globe's Central Urban Issues', 46(11), pp. 2259–2275; Watson, Plan Theory Pract 15:62–76, 2014b) argue that the diversity and uniqueness of each urban context developing at the intersection of local, regional and global challenges, threats and production of knowledge. In light of this, the chapter gives an insight into the conceptual framing of this book, including the key thematic areas; and an overview of topics covered by the chapters. The book has three thematic areas: planning theories and Models; the state of planning education and capacity; participatory and multi-governance approach towards current urban challenges. Under these themes, the chapter introduces several cases from various cities across Africa.

G. Alem Gebregiorgis (✉) · S. Greiving
International Planning Studies, TU Dortmund University, Dortmund, Germany
e-mail: genet.alem@tu-dortmund.de

S. Greiving
e-mail: stefan.greiving@tu-dortmund.de

A. H. Namangaya
School of Spatial Planning and Social Sciences, Ardhi University (ARU), Dar es Salaam, Tanzania
e-mail: ally.namangaya@aru.ac.tz

W. J. Kombe
Institute of Human Settlements Studies, Ardhi University (ARU), Dar es Salaam, Tanzania
e-mail: wilbard.kombe@aru.ac.tz

G. Alem Gebregiorgis et al. (eds.), *Planning Cities in Africa*, The Urban Book Series,
https://doi.org/10.1007/978-3-031-06550-7_1

Keywords Urbanisation · Planning Theories · Planning Models · Planning Education · Planning Practice · Multi-Level Governance

1.1 Background to the Concept of the Book: Planning Cities in Africa

Urbanisation and climate change were at the centre of our dialogue when the idea of this book was initiated. Its topics have been the focus of global and national policies and research such as the Sustainable Development Goals, for example, Goal 11 on Sustainable Cities and Communities, (United Nation 2015) and the New Urban Agenda (UN-HABITAT 2017), which vows to "commit … to strengthening the resilience of cities and human settlements, including through the development of quality infrastructure and spatial planning (…), especially in risk-prone areas of formal and informal settlements." The topics are linked with key local, regional and global factors posing a huge challenge to socioeconomic and natural systems and assets (Parnell and Walawege 2011; Addaney et al. 2017; Cobbinah et al. 2017; CRED and UNDRR 2020; Biesbroek et al. 2022).

Cities in the Global South are overwhelmed by an ever-growing demand for affordable housing and urban services. This is indeed a paramount challenge that African cities must deal with albeit their encumbered capacity. For instance, the absolute number of people living in urban areas has almost doubled, despite the fact that the proportion of residents living in slum settlements of African cities has slightly declined (UN-Habitat 2016). In cities such as Addis Ababa and Dar es Salaam, people living in slum or informal settlements account for more than 50 per cent of the total urban population (*ibid.*). This means that these cities must find a way to improve the substandard housing and sanitation conditions, and supply basic services for more than half of their population, while managing all other urban development issues related to environment, mobility, waste management and the like (Blanco et al. 2009; Parnell and Walawege 2011; Alem and Namangaya 2021). In addition to this dire situation, flood-related risks and the frequency and intensity of the resulting disasters are increasing. These are to some extent impacted by global climate change which causes erratic rainfall patterns and an increase in temperature (Addaney et al. 2017; Cobbinah et al. 2017; Addaney and Cobbinah 2019). However, the flood risk is a sum of these problems affecting the pattern of urban settlements, degradation of green and blue infrastructure and encumbered urban management capacity. Increasingly, informal settlements on hazardous and eco-sensitive areas are almost the norm in African urban landscapes and also in Latin America and Asia (Blanco et al. 2009; Greiving et al. 2018).

Concurrent to this, the body of literature and focus of discourses on urban studies are bringing new perspectives and approaches to understand urbanisation in Africa and sociocultural resources. These new perspectives and approaches enable an understanding of cities and urbanisation in Africa from their own sociocultural and historical perspective (Simone 2004, 2010; Fourchard 2011; Robinson 2011; Connell

2014). These approaches help to correct and update the Euro-American centric view that mostly sees African cities as synonymous with crisis and disaster (Fourchard 2011; Myers 2011; Robinson 2011). Despite the trends towards a change of perspective and calls to actualise and reform planning in Africa, in practice, there seems to be little change (Robinson 2006; Watson 2009, 2011, 2014b; Comaroff and Comaroff 2012; Cain 2014; De Satgé and Watson 2018). For the most part, planning education and practice in Africa are rooted in and originated from its colonial past. Theories, the ideal of "city" and urbaneness are hence derived from a Euro-American urbanism and sociocultural context (Diouf and Fredericks 2014; Wesely and Allen 2019). In fact, planning practice in the region is associated with rationalist master planning approaches and inappropriate strategies, which are developed for other sociocultural and economic contexts, in dealing with actual urban problems (Watson 2014a; Cirolia and Berrisford 2017; Ewing 2021). In several occasions, urban planning and urban development interventions are associated with injustice and marginalisation of the poor (Watson 2011; Alemu 2015; McClelland 2018; Alem 2021).

In view of this, the research team, under the four-year research-cum training programme on Governance and Planning of Resilient Cities in East Africa (GOPLAREA), developed three key thematic areas for a series of events which have become the background of this book.

1.2 Key Thematic Areas of GOPLAREA Research Project

The main thematic areas developed under the GOPLAREA research project are localising planning theories, innovation of planning education and participatory and multi-governance planning approach. The research and discourse on these topics were structured by several questions, which guided the research and development of concepts for building public dialogue and co-production of knowledge.

- **Localising planning theories**: What are the prospects of planning theory in the continent's urban planning landscape, which actual issues are important to address the conceptual gaps in planning practice and education?
- **Innovating planning education to address the combined challenges of rapid urbanisation and climate change**: How far have the curricula of urban planning education changed following the call for discourse on postcolonial planning theories and in particular the perspective from the south? Are there reforms on methods of knowledge acquisition and co-production? What were the impacts of increasing urban risks, and how have concepts such as sustainable development, urban resilience and risk adaptation? How have these influenced the teaching and training of planners?
- **Participatory and multi-level governance approach towards the combined challenges of rapid urbanisation and the impact of climate change**: Challenges posed by rapid urbanisation and climate change-related risks have further exacerbated the weak institutional and financial deficits of cities in Africa. On the

other hand, a considerable amount of resources are used in networks and initiatives in informal and pseudo formal systems. Among other things, these resources aim to address shortages of or inaccessible urban services, mitigation and adaptation to disaster-related risks. Some of the emerging questions are: What kind of resources are available and how best to deal with challenges in order to exploit opportunities of participatory and multi-levels governance approaches? How are resources at the local level mobilised and coordinated? How to coordinate and exploit such resources and make them effective for sustainable urban development and resilience building?

From 2016 to 2021, a series of workshops and a conference were organised by the research project. The workshops focused on discussions of critical issues of urban planning and the challenges of climate change impacts in East African cities. The aim of the workshops and conference was to create knowledge production platforms where researchers, practising planners and actors at the local and grassroots level could meet and discuss the burning issues in urban areas. In particular, the workshops facilitated to co-assessment, co-planning and administration needs of important planning and development agents. They also discussed issues, such as urban flood risks, ongoing initiatives of risk mapping and database building, nature-based solutions which could make African cities resilient, participatory management of urban river valleys and flooding and capacity building for resilient risk and urban management.

An international forum, which was organised as the fifth conference of AAPS (Association of African Planning Schools), brought together scholars from several countries in Africa and as well as international scholars. The conference had several parallel and plenary sessions, which facilitated opening the dialogue to a wide range of participants. More than 70 papers were presented and discussed in the three-day event. However, it is not the ambition of this book to summarise all the discussion of the papers presented and the keynote speeches, but to present a selection of topics, which provide a glimpse into the discussed topics in planning theory and education as well as current issues of urban governance and risk management.

1.3 Book Chapters and Key Thematic Areas

The book is organised in 11 chapters including the introduction. The chapters are organised in three thematic groups: Planning theories and models—application and local challenges; the state of planning education and planning capacity and; participatory and multi-level governance approach towards current urban challenges. The chapters cover several issues in climate change impact adaptation, participatory planning, co-production of services, housing and planning education. Implications of Euro-American centred planning models, concepts and theories are also critically assessed, as well as the prospects for contextualisation and local perspectives of urban challenges.

In Part I, Lia Gebremariam Woldetsadik, Wafae Boullala and Raffael Beier, the authors question, in different ways, the application of imported planning theories, models and concepts. They have sought to highlight the implication of theories and concepts that draw ideals from foreign sociocultural and economic contexts and experiences. Woldestadik, taking the case of neighbourhood upgrading initiatives in Addis Ababa, discusses the application of collaborative planning and how it is affected by government systems. She notes that democratic culture, trust and cooperative norms are linked to government systems shaping the planning process and cooperation. Hence, the state, in dominantly authoritarian regimes, penetrates the society and shaping social norms and the worldview of actors. She argues that the collaborative planning process should be designed taking into consideration contextualised state-society relationships. Wafae Boullala, in Chap. 3, thematises the issue of street vending and the modernist view of "informality" using historical research in Rabat. She notes that imported planning models have led to the marginalisation of people, who, as peddlers, were part of a centuries-old tradition of the "*Souk.*" Boullala argues that street vending is an economic activity linked to the historical development of Rabat, which has been pushed out of the downtown by colonial planning models and urban policies. In her view, street vending is a form of resistance against alien policies and urban forms; and a struggle to claim back urban spaces. Raffael Beier, in his chapter titled "Revisiting Stokes Theory of Slums," discusses the concept of the slum and its implications in the development of large-scale housing programmes. Using cases of different housing programmes in various African countries, he analyses the conceptual underpinning behind adequate housing. Beier notes of the implications of conceptualisation that neglects social capital embedded in the communities living in informal settlements and the social entrepreneurship in self-help housing development. He argues that concepts such as "slums" are developed from colonial planning and mostly concerned with the notion of shelter. Hence, he calls for a contextualised concept of adequate housing which goes beyond material decency.

The chapters in Part II discuss the state of planning education and planning capacity in different planning systems in Africa. Chapters 5 and 6 discuss the education and training of planners, analysing the content of course modules. Geetika Anand and Nadini Dutta, Chap. 5, and Abraham Matamanda, Jennilee Kohima, Veran Nel and Innocent Chirisa, Chap. 6, present assessments of planning education curricula and how planners are trained and equipped to deal with contemporaneous urban challenges. Geetika Anand and Nadini Dutta present us a comparative study of different planning curricula in ten universities in Africa and Asia. The chapter is also a reflection on the state of planning education in a postcolonial context. The authors note that the reviewed planning curricula exhibit a dependency on education philosophy, pedagogy and literature, which are mainly derived from Euro-American experiences. In addition, current issues, such as informality, access to land, sustainability and justice remain either on the periphery or at a very abstract level. Abraham Matamanda et al., also review curricula, but focus on to what extent knowledge of the implications of climate change and adaptation are taught in planning schools of South Africa,

Zimbabwe and Namibia. The authors argue that the contents show acknowledgement of the impacts of climate change, but it fails to include teaching module which deal with contextualised challenges, such as flooding and draught.

In contrast, Wilbard Kombe and Samwel Alananga and Dawah Magembe-Mushi and Ramadhan Matingas, analyse the capacity and skills of planners and officials on the ground. The authors assess the capacity and skills of planners and officials of local authorities in managing planning processes and understanding issues related to climate change. Kombe and Alananga, Chap. 7, discuss the link between the knowledge that planning professionals have and the achieved urban resilience in Tanzania. Assessing the case of Arusha Municipality in Tanzania, they observe the capacity gaps related to the training and education of professional planners. The authors argue that education and awareness on climate change impacts are key pillars to meet the aims of climate change-related policies at local level. Magembe-Mushi and Matingas, Chap. 8, take the case of sea level rise in the Pangani Divison, in Tanzania, and assess the role of authorities in working with the community and implementing localised adaptive measures. The authors underscore the role of capacity building to equip local authorities with adequate leadership quality and skills in managing initiatives in climate change adaptation and mitigation. In addition, a conducive information sharing environment and efficient resource mobilisation contribute to community resilience building.

In Part III, three chapters discuss the landscapes of participatory planning and multi-level governance using the cases of planning process in Bahir Dar, in Ethiopia and co-production of urban spaces in Maputo, in Mozambique, and infrastructure in Wa, in Ghana. In Chap. 9, Behailu Melesse Digafe, Achamyeleh Gashu Adam, Gebeyehu Belay Shibeshi and Mengiste Abate Meshesha present the process through which the Structure Plan of Bahir Dar, in Ethiopia, was prepared. The authors underline a conflict emerging from combining participatory planning process with the top-down planning, i.e. pre-set planning standards. They argue that stakeholder participation and acceptance of plans by the community could be improved if planning standards were flexible and limited to a general framework guiding the participatory planning process. The authors call for due attention to communal interests and local priorities, while making information on important planning requirements available to the communities and stakeholders in general.

Milousa Ibraimo António, Chap. 10, and Francis Dakyaga, Chap. 11, bring the issue of participatory and multi-level governance to the fore. António examines self-production practices within the context of the production of collective spaces. Using the concept of alternative informality, she discusses how the incremental housing construction at the household level is linked to community level informal land administration practices. Using the case of a neighbourhood at the periphery of Maputo, she notes the interplay of different levels of governance with informal ones at the grassroots level. The chapter underlines that there is a wide consensus that makes local leaders, informal actors of land governance, responsible for managing and guiding the production of collective spaces. In Chap. 11, Dakyaga discusses post settlement planning and development of infrastructure for clean water supply. Using the networked city concept, the chapter explores the co-production of infrastructure and management of clean water supply systems in different level of governance. He notes that the heterogeneity of supply models is essential to support the effort towards universal water access. Dakyaga argues that collaboration among the different types of actors is decisive for filling the institutional gap in supplying basic services.

References

Addaney M, Boshoff E, Olutola B (2017) The climate change and human rights nexus in Africa. Amsterdam Law Forum 9(3):5. https://doi.org/10.37974/ALF.304

Addaney M, Cobbinah PB (2019) Climate change, urban planning and sustainable development in africa: the difference worth appreciating. In: Cobbinah PB, Addaney M (eds) The geography of climate change adaptation in urban Africa. Springer, Cham, pp 3–26. https://doi.org/10.1007/978-3-030-04873-0_1

Alem G (2021) Urban plans and conflicting interests in sustainable cross-boundary land governance, the case of national urban and regional plans in Ethiopia. Sustain 13(6):3081. https://doi.org/10.3390/su13063081

Alem G, Namangaya AH (2021) 'Households' and community initiatives towards city resilience: the case of flood resilience in Dar es Salaam. In Faldi G, Fisher A, Moretto L (eds) African cities through local eyes: experiments in place-based planning and design. Available at: https://doi.org/10.1007/978-3-030-84906-1. Accessed 14 Jan 2022

Alemu GTe (2015) Land expropriation and compensation payment in Ethiopia. Review 6(13):93–97

Biesbroek R et al (2022) Policy attention to climate change impacts, adaptation and vulnerability: a global assessment of national communications (1994–2019). Clim Policy 22(1):97–111. https://doi.org/10.1080/14693062.2021.2018986

Blanco H et al (2009) Shaken, shrinking, hot, impoverished and informal: emerging research agendas in planning. Prog Plan 72(4):195–250. https://doi.org/10.1016/j.progress.2009.09.001

Cain A (2014) African urban fantasies: past lessons and emerging realities. Environ Urban 26(2):561–567. https://doi.org/10.1177/0956247814526544

Cirolia LR, Berrisford S (2017) "Negotiated planning": diverse trajectories of implementation in Nairobi, Addis Ababa, and Harare. Habitat Int 59: 71–79. https://doi.org/10.1016/j.habitatint.2016.11.005

Cobbinah PB, Poku-Boansi M, Peprah C (2017) Urban environmental problems in Ghana. Environ Dev 23:33–46. https://doi.org/10.1016/j.envdev.2017.05.001

Comaroff J, Comaroff JL (2012) Theory from the south: or, how Euro-America is evolving toward Africa. Anthropol Forum 22(2):113–131. https://doi.org/10.1080/00664677.2012.694169

Connell R (2014) Using southern theory: decolonizing social thought in theory, research and application. Plan Theory 13(2):210–223. https://doi.org/10.1177/1473095213499216

CRED, UNDRR (2020) Human cost of disasters: an overview of the last 20 years, 2000–2019. UNDRR, p. 30. Available at: https://www.preventionweb.net/files/74124_humancostofdisasters20002019reportu.pdf. Accessed 7 July 2021

De Satgé R, Watson V (2018) Urban planning in the global south: conflicting rationalities in contested urban space. Springer, Palgrave Macmillan, Cham. Available at: https://doi.org/10.1007/978-3-319-69496-2. Accessed 1 Nov 2020

Diouf M, Fredericks R (2014) Introduction: decolonizing urban theory. In Diouf M, Fredericks R (eds) The arts of citizenship in African cities. Palgrave Macmillan US, New York, pp 1–23. https://doi.org/10.1057/9781137481887

Ewing K (2021) Spaces of transformative practice: co-producing, (re)making and translating fractional urban space in Gugulethu, Cape Town. Urban Forum 32(4):395–413. https://doi.org/10.1007/s12132-021-09436-6

Fourchard L (2011) Between world history and state formation: new perspectives on Africa's cities. J Afr Hist 52(2):223–248. Available at: https://www.jstor.org/stable/23017677. Accessed: 2 July 2020
Greiving S, Du J, Puntub W (2018) Managed retreat—A strategy for the mitigation of disaster risks with international and comparative perspectives. J Extreme Events 5(2/3):1850011. https://doi.org/10.1142/S2345737618500112
McClelland J (2018) Planners and the work of renewal in Addis Ababa: developmental state, urbanizing society. PhD. University of Washington. Available at: https://digital.lib.washington.edu/researchworks/bitstream/handle/1773/41805/McClelland_washington_0250E_18306.pdf?sequence=1&isAllowed=y. Accessed 29 Jan 2021
Myers GA (2011) African cities: alternative visions of urban theory and practice.Zed Books Ltd., London and New York
Parnell S, Walawege R (2011) Sub-Saharan African urbanisation and global environmental change. Glob Environ Chang 21:S12–S20. https://doi.org/10.1016/j.gloenvcha.2011.09.014
Robinson J (2006) Ordinary cities: between modernity and development.Routledge (Questioning cities), London and New York
Robinson J (2011) Cities in a world of cities: the comparative gesture: cities in a world of cities compared. Int J Urban Reg Res 35(1):1–23. https://doi.org/10.1111/j.1468-2427.2010.00982.x
Simone A (2004) People as infrastructure: intersecting fragments in Johannesburg. Publ Cult 16(3):407–429. https://doi.org/10.1215/08992363-16-3-407
Simone A (2010) City life from Jakarta to Dakar: movements at the crossroads. Routledge. https://doi.org/10.4324/9780203892497
UN-Habitat (2016) Urbanization and development: emerging futures. UN-Habitat (World cities report, 2016), Nairobi, Kenya
UN-HABITAT (2017) New urban agenda: Habitat III: Quito 17–20 October 2016. United Nations Conference on Housing and Sustainable Urban Development. United Nations, Nairobi
United Nation (2015) Transforming our world: the 2030 agenda for sustainable development. United Nations. Available at: https://sustainabledevelopment.un.org/
Watson V (2009) Seeing from the south: refocusing urban planning on the globe's central urban issues. Urban Stud 46(11):2259–2275
Watson V (2011) Changing planning law in Africa: an introduction to the issue. Urban Forum 22(3):203–208. https://doi.org/10.1007/s12132-011-9118-9
Watson V (2014a) African urban fantasies: dreams or nightmares? Environ Urban 26(1):215–231. https://doi.org/10.1177/0956247813513705
Watson V (2014b) Co-production and collaboration in planning—The difference. Plan Theory Pract 15(1):62–76. https://doi.org/10.1080/14649357.2013.866266
Wesely J, Allen A (2019) De-Colonising planning education? Exploring the geographies of urban planning education networks. Urban Plan 4(4):139–151. https://doi.org/10.17645/up.v4i4.2200

Dr. Genet Alem Gebregiorgis is a lecturer in the Department of International Planning Studies of the Faculty of Spatial Planning at TU Dortmund University. She has worked as a lecturer at Dohuk University in Iraq and at the Ethiopian Civil Service University in the Urban Management Masters' Programme. Genet Alem also has several years of experience practicing as a professional architect and planner. In Ethiopia, she has worked for the Federal Urban Planning Institute, Addis Ababa Works and Urban Development Bureau, and Office for the revision of the Addis Ababa Master Plan. Alem holds a Ph.D. in Spatial Planning from TU Dortmund, an MSc in Urban and Regional Planning from TU Dortmund and Dar es Salaam University, and a degree in architecture from The Technical University of Havana José Antonio Echeverría (La CUJAE) in Havana, Cuba. Genet Alem is member of Network-Association of European Researchers on Urbanization in the South (N-earus), Planners for Climate Action (P4CA) and TRIALOG e.V.—Association for Scientific Research into Planning and Building in the Developing World.

Prof. Ally Namangaya has been in practice as a land use planner since he completed his training. He is an associate professor at Ardhi University. He obtained his PhD at TU Dortmund University in 2010. He is currently Dean in the School of Spatial Planning and Social Sciences. He has undertaken research and been published in peer-reviewed journals on issues such as urban competitiveness and land-based financing; rural land use conflicts; community based natural resource management; the application of GIS in land use planning; and the linking of rural communities to tourism value chains. He has extensive experience as a consultant and in research on the reforms of urban planning practices and the incorporation of digital technologies into urban planning. He has led the preparation of master plans for Kibaha, Geita and Morogoro, and several neighbourhood and institutional master plans (estate development plans). On rural development planning, he has worked on community engagement and land use planning for out-grower development project entities such as RMI, Eid environment, Eco-Energy and Southern Agricultural Growth Corridor of Tanzania (SAGCOT). He is the non-executive founding Director of the PAMS Foundation, which is dealing with anti-poaching and conservation in the Selous Niassa Wildlife Corridor and the Ngorongoro Conservation Area and its surroundings.

Prof. Dr. Stefan Greiving has been head of the Institute of Spatial Planning at TU Dortmund University since 2011. He holds a degree in spatial planning and a Ph.D. in urban planning. His postdoc thesis (habilitation) focused on spatial planning and risk. Prof. Greiving has been PI of about 40 international and 80 national projects on disaster risk management, adaptation to climate change, and regional development. He is the author of about 250 publications. Prof. Greiving is a member of the German Academy for Territorial Development and a member of the chamber of architects.

Wilbard Jackson Kombe is a professor at the Institute of Human Settlement Studies at Ardhi University in Tanzania. Kombe received his Ph.D. from the Technical University of Dortmund. Since his graduation, he has extensively collaborated with various institutions, both locally and internationally, in research and training activities. His recent research work has focused on land management and urban governance, informal urbanisation in poverty, urbanisation, and public policy analysis. He has been extensively involved in applied and basic research activities in numerous countries in Sub-Saharan Africa. This includes research activities undertaken in Tanzania, South Africa, Zimbabwe, Kenya, Uganda, Mozambique, Zambia, Namibia, Botswana, and Ethiopia. He has also led a number of multi-disciplinary research teams and projects undertaken in Tanzania and other countries in the region.

Part I
Planning Theories and Models—Application and Local Challenges

Chapter 2
The State, Trust and Cooperation: Local Government-Residents' Joint Neighbourhood Upgrading Initiatives in Addis Ababa

Lia Gabremariam Woldetsadik

Abstract Collaborative planning is acknowledged to enable several positive outcomes including the building of local knowledge and capacities. It is deemed to facilitate mobilisation of resources, support, acceptance, coordination of action and ownership. However, the collaborative planning literature's focus on techniques to perfect the process design (i.e., regarding modalities for structuring participation, communication and deliberations) to transform social and political institutions pays little attention to the penetration of planning practices by the overall institutional environment that impede or enable operationalising these techniques. And based on Western liberal realities, it presumes that a minimum level of trust and at least democratic culture and cooperative norms needed for collaborative planning exist everywhere. As de Satgé and Watson (Urban planning in the global south: conflicting rationalities in contested urban space, Springer, 2018) argue, the "thin and instrumental assumptions" that planning theories make regarding the applicability of public participation or collaborative planning do not fit in with what is on the ground in other contexts, such as what is found in many parts of Africa. The chapter aims to bring the state back into the collaborative planning discourse by analysing how government systems affect the conceptions and actions of the different urban actors in collective action. Through local government-residents' joint urban upgrading projects in two localities of Addis Ababa, it provides insight into the link between government systems, trust, planning practices and cooperation.

Keywords Collaborative planning · Collective action · Government systems · Trust · Cooperation

---

L. G. Woldetsadik (✉)
Ethiopian Institute of Architecture, Building Construction and City Development, Addis Ababa, Ethiopia
e-mail: lia.gmariam@eiabc.edu.et

G. Alem Gebregiorgis et al. (eds.), *Planning Cities in Africa*, The Urban Book Series,
https://doi.org/10.1007/978-3-031-06550-7_2

2.1 Introduction

More democratic planning processes are acknowledged to facilitate several positive outcomes, including acceptance, support, better solutions and ownership (Healey 1998; Hillier 2002; Innes and Booher 2004; Albrechts 2013), which are even more critical in developing countries where resource limitation is debilitating. The paradox lies in needing to institute more democratic planning processes inside less democratic systems. Many external variables influence the process design as planning and its governance model evolve and operate inside a system context where political dynamics, power relations and other related institutional characteristics exist (Huxley and Yiftachel 2000; Fainstein 2005; Brand and Gaffikin 2007).

Even though collaborative planning aims to transform social and political institutions by building shared meaning, civic capacity or capable civil society, social capital and more participatory forms of democracy (Healey 1998; Innes and Booher 2004; Booher and Innes 2005), its focus on operationalising certain techniques in the process design for neutralising power imbalances and building consensus pays little attention to factors in the overall system context that can hinder making these possible. It takes for granted that the conditions for creating democratic planning processes and rigorous discourse are always readily available everywhere. For instance, the role of the planner in the collaborative discourse is to empower citizens (Forester 1987), facilitate consensus building (Forester 1987; Healey 1998) and social learning (Booher and Innes 2005), build trust (Emerson and Gerlak 2014) and provide collaborative leadership by ensuring the integrity of the consensus building process (Ansell and Gash 2008). But planners are not always "open to democratic processes" (Tewdwr-Jones and Allmendinger 1998: 1984). They are neither neutral nor can remain apolitical experts and cannot be entirely free from personal motivation or pre-packaged values and preferences born out of sociopolitical and economic contexts (Tewdwr-Jones 2002).

As a source of institutional order with inherent characteristics that produce asymmetry in power relations (O'Donnell 1993), it is the state that shapes society (Skocpol 1985), and the nature of the state and government systems influence the system context in which planning is carried out. This influence is not limited to its implication on the planning process, as planning is largely a public domain that is undertaken "by, or in relation to the state" (Huxley and Yiftachel 2000: 106), also requiring transparency and accountability in government transactions (Albrechts 2004). The influence of government systems on planning extends well beyond this. For instance, since democratic systems allow more oversight of governments (Delhey and Newton 2004) and ensure their accountability (O'Donnell 1993), instituting collaborative or democratic planning processes requires a democratic system context because organisational practices and structures are the reflection of the overall institutional environment (Powell and DiMaggio 1991). Similarly, the importance of trust is a widely discussed input and outcome norm in the collaborative literature. Collaborative or participatory planning processes are said to build trust among participants, while social capital is expected to enable participants to generate and share knowledge, create understanding and build trust as well as mobilise participants into action (Healey 1998, 1999). The proponents of collaborative/participatory planning and collaborative governance argue that continuous interactions are expected to build trust and stronger relations among participants, as trust ensures stakeholder commitment to the process (Ansell and Gash 2008). This seems to emphasise particularised trust as an input norm, which limits the expectation of reciprocity to only those involved, as opposed to social trust which assumes that people who are not kin, close associates, or who do not have shared experiences can generally be depended upon to act for the common good. On the other hand, an initial basic level of trust needs to exist in society (Brand and Gaffikin 2007) in order to enable the planning process to build trust among participants. This also includes trust for process organisers (Newig et al. 2018).

Collaborative planning or any other form of democratic planning process also necessitates some level of democratic culture in society for both process organisers and participants to effectively negotiate or diffuse power and cooperate in collective action. Democracy also encourages the development of social belief systems such as social trust (Uslaner 1999; Delhey and Newton 2004). In contrast, undemocratic regimes are neither consensual nor power sharing (Freitag and Bühlmann 2009). They destroy the development of alternative sources of power such as civil society (Uslaner 1999) and lack institutions that promote wider political engagement (Bratton and Van De Walle 1994), thereby eroding the development of democratic culture (Agger and Löfgren 2008) and confidence or trust in government. Lack of trust or confidence in government, in turn, inhibits the development of social trust and cooperation (Rothstein and Teorell 2008; Freitag and Traunmüller 2009; Robbins 2011). Cooperation is the we-intention and joint action (Tuomela and Miller 1985), or a set of actions taken by two or more individuals to achieve the same or complementary goals based on interdependence (Deutsch 1949) or reciprocity, which may require some measure of sacrifice (Khamis et al. 2006) built on individuals' moral value (Daniel 1994) or trust (Deutsch 1949).

Through local government-residents' joint neighbourhood upgrading projects in Addis Ababa, the article shows the extent to which lack of democratic culture, trust and cooperative norms are closely linked with government systems, also shaping planning processes and cooperation. The article investigates why residents in two localities of Addis Ababa choose to cooperate or not in joint neighbourhood upgrading projects that local governments convene. The article aims to bring the state back into the collaborative planning discourse by analysing how government systems affect the conceptions and actions of the different urban actors in collective action.

2.2 Methodology

The research primarily adopts the qualitative approach and uses interviews and surveys to analyse the dynamics of collective action in two districts in Addis Ababa through local government-residents' joint urban upgrading projects. The discussion investigates relations between government systems, planning processes, trust and cooperation in the dilapidated poor inner-city old neighbourhoods of the capital, and in poorly serviced "informal" and relatively new settlements in expansion areas of Addis Ababa.

Interviews with semi-informal resident development committee members and local government actors are used to investigate structures, processes, acceptance, support and ownership in the joint initiative. The parameters acceptance, support and ownership are used to investigate the state of cooperation. Acceptance is assessed based on favourable reception of spatial policy goals/projects. While resident support for the joint initiative is expressed in the degree of voluntary involvement and resource mobilisation, ownership is assessed in terms of the management of communal facilities and public infrastructure built through the joint initiative by beneficiaries. And through a survey of randomly selected households, the research determines the level of confidence in the respective local governments by using two parameters, corruption and the quality of service delivery in the local government, asking respondents to grade the levels of each parameter for confidence in the local government on a scale from zero to four, representing very low, low, moderate, high and very high, respectively.

By using the same grading system, the research establishes public confidence in resident development committees. To determine the type of trust that exists in the respective areas, the survey asked respondents if they sometimes leave their houses unlocked. Interviews with randomly selected households were used to establish if residents in the two localities consider that people can be generally trusted to determine the level of social trust in the respective localities. The survey also included queries designed to determine the extent of participation and transparency in planning, accountability, acceptance, support, ownership and trust.

2.3 Background

The nationalisation of extra urban houses through proclamation 47/75 in 1975 has had a negative impact on the physical environment of the capital of Ethiopia, Addis Ababa. Depending on the amount of monthly rents, the administrations of the nationalised houses were transferred to either *kebele*[1] administrations or to a government agency established for this purpose. Most *kebele* houses are mud-constructed attached

[1] A *kebele* was the lowest formal local government structure in Addis Ababa until recently. It was replaced by "*woreda*" following the reorganisation of the Addis Ababa city administration which had merged, in some cases, two or more *kebeles* into one *kebele*.

single-story single-room tenements. In 2008, they represented up to 66% of residential houses in the inner-city, where 42% of the city's residents were living on only 11.2% of the total built-up area (Yitbarek 2008). The fact that nearly 82% were rented out for less than 30 birr (less than €2) per month had made the generation of funds for their maintenance difficult (Yitbarek 2008). Moreover, the 1986 Master Plan of Addis Ababa that prohibited the upgrading of such houses had prevented the able and willing tenants from maintaining their homes for more than two decades. Little has changed regarding the ownership structure and predominance of *kebele* houses after more than four decades, while the value of money collected from rent continues to decline and the state of dilapidation worsens. Moreover, sub-division and sub-letting by tenants despite restrictions on unauthorised transformations, as well as widespread informal land allocation by *kebele* authorities, has not only perpetuated illegal housing construction in the inner-city (UN-Habitat 2010), but also intensified overcrowding and the physical deterioration of most of the older inner-city neighbourhoods.

On the other hand, as the sole owner and administrator of land in Addis Ababa, the city government has neither been able to adequately supply serviced land for housing nor effectively manage lateral irregularities in land administration. The introduction of the urban land lease system in 1994 initiated the practice of attaching market value for the use right of urban land and allocating plots based on auctions. Given the limited supply of land, the system has pushed land prices up, and out of reach for the majority. As a result, a significant number of low and middle-income residents have resorted to illegally invading urban land and making unauthorised purchases from farmers on the periphery. For instance, in the six years between 1985 and 1991, the number of new, unauthorised houses built on illegally occupied land reached 13,104, but by 2002 the total number exceeded 60,000 (ORAAMP 2002). Even though government intervention regarding squatter settlements mainly constitutes bulldozing, those who had informally settled on the fringes before 1995 (and later before 2005) were legalised provided the land was not immediately needed for "public purposes", or their safety and the safety of others was not endangered. The layouts of some of these settlements are better organised, and the houses, in some cases, are built from more durable construction materials than found in many inner-city neighbourhoods. Squatting and illegal purchases has continued to grow, and bulldozing is sometimes met with violent resistance.

The Public Participation Development Agency (PPDA) was established in 2011 under the Addis Ababa City Government's Construction and Housing Development Bureau[2] to facilitate collaboration between the city government and residents in neighbourhood upgrading. In addition to physical infrastructure and communal facility maintenance and construction, the PPDA is also responsible for coordinating, supporting and monitoring public participation in the maintenance of houses for "the poorest of the poor". It operates through the Public Participation Development

[2] Due to reorganisation of public offices, the PPDA is now under the Housing Administration Office, which is directly accountable to the city's Mayor.

Offices (PPDOs), setup at the existing 10 sub-cities (*Kifle Ketemas*)[3] and 116 *woreda* administrations. Overall, 2.04 billion birr (around €73 million) worth of support was mobilised between 2011 and 2015 mainly from residents through the different development committees established at woreda and zone[4] levels. In 2016, some 1,044 residents were organised in woreda core development committees (9 members per woreda), and 11,580 in zone development committees (15 members per zone). The joint initiative also uses development teams in "menders"[5] and the one-for-five structure[6] for mobilising residents. An annual budget of 100 million birr (€1.97 million at the current exchange rate) is allocated by the city government for this purpose; and a flat rate of 65% matching fund needs to be mobilised from the community by the respective development committees and deposited in a bank account dedicated for this purpose before the government's share of 35% for the respective projects is released.

Residents elect woreda and zone level development committees through a general assembly, and members provide their services from their free time without payment. These development committees, in collaboration with the respective woreda administrations, plan, collect community contributions and facilitate the implementation of the neighbourhood upgrading projects. Woreda PPDOs support activities of zone development committees and integrate zone committees' plans by prioritising projects based on pledged contributions in the respective zones; and woreda level core development committees submit their annual plan and copies of bank deposit receipts to the sub-city PPDO, upon which the sub-city PPDO writes a letter to Addis Ababa City Government Bureau of Finance and Economic Development for the release of government matching funds. Interviewees from the PPDA's wing responsible for planning and public mobilisation processes contend that the programme intentionally promotes the involvement of residents in the planning, implementation, monitoring and management of the public goods and communal facilities built through the initiative to enhance community ownership.

[3] Bole, Yeka, Lideta, Arada, Gullele, Kirkos, Nefas-silk, Akaki-kaliti, Kolefe-keraniyo and Addis Ketema sub-cities. Since October 2020, Lemi Kura has become the eleventh sub-city by incorporating and combining parts of Yeka and Bole sub-cities.

[4] The zone structure in Addis Ababa is not a formal administrative structure and covers a narrower geographic area than the woreda.

[5] Mender or Sefer is not a formal administrative structure and refers to a neighbourhood that covers a narrower geographic area than the zone. There are usually up to seven menders in each zone.

[6] Although there were some allegations that the one-for-five structure was being used by the government to increase party membership (https://en.wikipedia.org/wiki/Hailemariam_Desalegn) and political control (http://www.theguardian.com/global-development/2015/may/22/ethiopia-elections-controlled-political-participation), it nonetheless is a pyramid-like configuration where one person is responsible for the development-related activities of five others.

2.4 Addis Ketema Sub-City, Woreda 6

Woreda 6 of Addis Ketema sub-city is located inside the deteriorated inner parts of Addis Ababa. There are around 38,000 residents in the woreda; and the size of the woreda is 64 ha. The survey was conducted on a randomly selected 154 households with an average family size of 5.9. According to the PPDA, the sub-city was the poorest performer in 2016 in terms of the volume of physical upgrading work it undertook and the resources it mobilised from residents for the joint neighbourhood upgrading initiative. The woreda development council, which is chaired by the Woreda 6 office of the ruling party, comprises 178 members of woreda and zone development committees, NGOs operating in the woreda, businessmen, community elders and religious leaders, as well as representatives of the local government.

The local government estimates that there are just over 1,500 housing units in the two zones selected for the research. Most of these housing units do not have a private kitchen or toilet and use ditches for sewage. These two zones were selected based on their performance with respect to resource mobilisation and asset management, where the performance of one was considered the best and the other the poorest by officials of the woreda administration. However, analysis of the survey data indicates that there is no significant difference between the responses of the randomly selected residents from the two zones regarding trust perceptions on processes and outcomes, and confidence in the local government and development committees. The only observable difference lies in the physical state of public goods (local cobblestone roads and drainages) and facilities (communal toilets and kitchens) built through the joint initiative. While the public goods and communal facilities in some areas of the better performing zone are in good condition, most built through the collaborative initiative in the poorer performing one are either partly damaged or no longer functional. The open ditch constructed along cobblestone roads for drainage in the poorer performing zone is clogged with debris and dirt. There are several spots where household wastewater has collected and overflows onto the road, which becomes worse during the rainy season. Some of the stones have been removed from the road and communal toilets in some areas no longer function (Fig. 2.1).

Even though there are differences in the average income level of households at zone or even mender levels, out of 154 households randomly selected for the survey, less than half (41%) state that they did not make financial contributions to the joint initiative because they were poor. Depending on the specific project it was intended for, households in the respective zones had contributed around 300 birr (less than €6 at the current exchange rate) annually. Almost three-quarters of those surveyed (72%) state that the amount each household contributed for the different projects of the local government-residents' joint upgrading initiative was either decided by semi-informal development committees established at zone levels or imposed by the woreda administration. This indicates that contributions were not voluntary as was claimed by officials. In addition, the fact that 83% of the respondents of the survey express high dissatisfaction with the physical outcome and 72% with the overall process can be evidence that the process was not participatory.

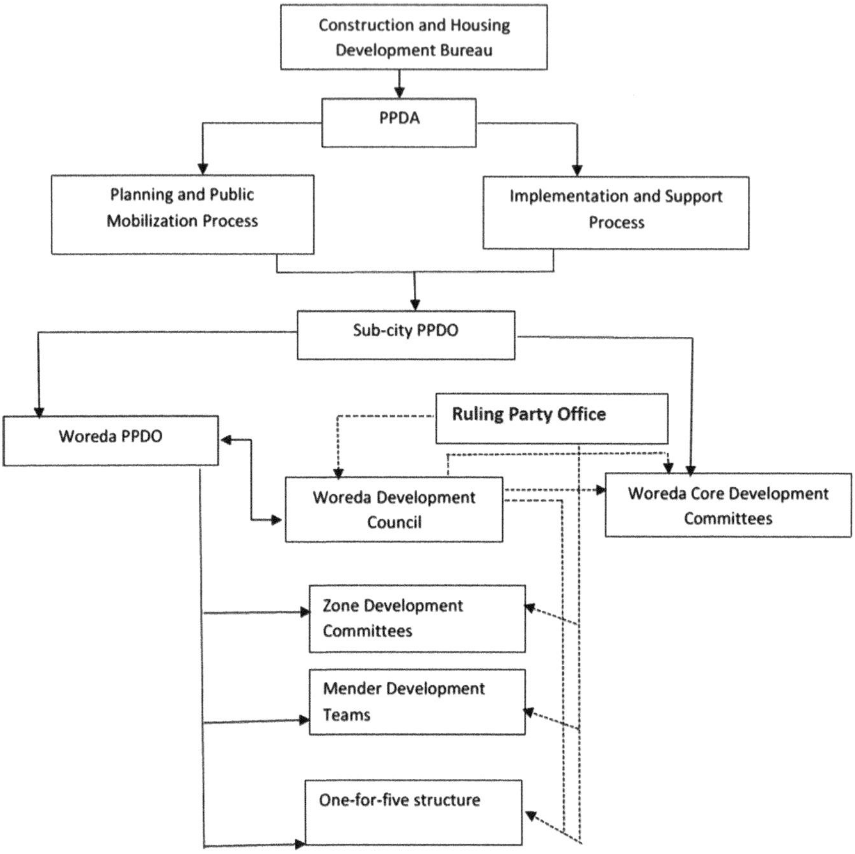

Fig. 2.1 Structure of a local government collaborative neighbourhood upgrading programme in Addis Ababa (*Source* Interviews with experts from the PPDA and PPDOs, and members of development committees)

On the other hand, members of zone and woreda core development committees argue that the minimum amount regarding specific projects is decided collectively through a general assembly. However, they also state that since only a few appear for these meetings, most residents do not have firsthand knowledge. Members of zone level development committees contend that when an amount is proposed by the floor, the motion usually passes unanimously because anyone who dares to oppose is labelled "anti-development". One adds,

> We are silent flunkies. We accept and obey any decision that the local government makes without questioning for fear of repercussions. We are even afraid to resign from the committee. The people are in the same boat. So, they have no choice but to agree. They think we are government reps. Then the problem becomes ours because they won't give us the money when we go to their doors to collect.

Another adds,

> They give us a couple of hundred birr and expect us to provide the roads and facilities at each mender. We have to select what to do and where. Besides, everyone is generally tired of contributing for several causes and asks us why the government does not do it on its own.

He further elucidates that this is because local government actors repeatedly inform residents that they are required to make contributions not because the government does not have the money, but so that the residents will "own" the projects and products.

The survey shows that confidence in government and the semi-informal development committees is low. While 69% responded that their confidence in the semi-informal development committees is between low and very low, at least three-quarters of the respondents have little confidence in the local government (woreda administration). Indicative of this, 77% claim that the quality of service delivery by the woreda administration is between low and very low, and 75% claim that the level of corruption in the woreda administration is between high and very high. Furthermore, 34% claim that they either refrained from making financial contributions or contributed less than what was required because they did not believe that the money would be used for the intended purpose, while 21% claim that they only attend meetings convened by local government actors because they believe that they can get some sort of benefit by being in the good graces of local government actors.

Two individuals who had served as members of zone committees in different capacities for several years argue that mismanagement of resources mainly occurs during procurement and supervision. One of these individuals says that he had left the committee of his own volition because he could not stop local government actors from claiming purchases they did not make, awarding contracts through illicit agreements and endorsing low quality products and services. Another claims that local government actors force the hands of those in committees into authorising payments for low quality physical work, and associates lack of trust by residents for the committees with corrupt practices in the local government and delays due to civil service lethargy. He says,

> The people want to see the finished products overnight and think that we have taken their money because of delays. Engineers in the woreda administrations are supposed to supervise the constructions. We are volunteers, we serve in our free time and we need government support every step of the way. They [government actors] get paid but just sit there doing nothing, and everything takes a long time before we can make headway.

Other committee members share these sentiments.

A member of one zone development committee explains that prior to 2015, activities of zone and woreda committees were not audited and thus, some members of zone and woreda committees have misappropriated resident contributions. He elaborates,

> Due to lack of transparency in the local government, the people do not know how much money was embezzled. They probably think it is larger than it is and assume that we are doing the same thing. These individuals were not held accountable. In fact, one individual who was also a member of the Addis Ababa City Council was reassigned to a better position. This is the system we have in Ethiopia.

This individual was a member of the ruling party, and such individuals are perceived, and more often treated as untouchables.

Most respondents argue that it is only a few culprits who misuse facilities and selfishly cause damages to the roads constructed through the joint initiative. Even though 59% state that they have either notified the committees or the woreda (17%), 62% suggest that no action was taken. But officials in the local government and woreda PPDO disagree with this and argue that people do not intervene or report to the woreda administration when infrastructure and facilities are harmed. "They rather stay silent than quarrel with neighbours they have eaten with". In Amharic, the phrase "they have eaten with" implies "those who have been there with you through thick and thin". This attitude was also suggested by a senior official and an expert working in the PPDA. Nonetheless, the state of the physical infrastructure and facilities shows that lack of conformism/compliance, and confidence in the development committees and the local government are also perpetuated by a lack of accountability. This confirms Ostrom's (1986) argument that ensuring accountability is paramount for collective action and the assertion by Khamis et al. (2006) that states that sanctions and enforcement are necessary for cooperation. Related to this, 12% contend that they refrained from making financial contributions because the infrastructure or facilities built through such endeavours would be damaged soon enough.

Most of those interviewed indicate that people generally are not to be trusted. However, 70% of the respondents maintain that they sometimes leave their doors unlocked, which may indicate that they trust their neighbours. This could also be related to the settlement pattern and housing typology as well as shared facilities that enable more interactions than in areas where secluded and more private living is possible. On the other hand, interviews with members of development committees provide a different perspective on social capital that exists in the locality. For instance, one member of a zone development committee contends,

> Let us say we plan to build or repair an existing communal toilet or kitchen for ten households living inside a compound and ask each to contribute a flat rate. Some refuse to make equal contribution by arguing that they have smaller family sizes than their neighbours. Others who are not direct beneficiaries of the communal toilets urinate on the walls. And we had to spend nights guarding the dumped sub-base material and cobblestones because some residents were removing these materials to lay the ground inside their compounds or homes.

They also contend that people contribute only if they directly benefit from the intervention. And even then, the resource they willingly contribute depends on their calculation of the relative benefits to others. Another committee member suggests that some of the residents still perceive these facilities and public goods produced through the joint initiative as government property. He argues, "They are unhappy with the political system and think that they are stealing from the government". Another agrees and states, "Some do not trust or like the government. Refusal to contribute and misuse or causing intentional damage is a reaction against the (government) system".

Even though there are several informal community associations such as *iddirs* in the area, which most residents are members of, the type of social capital or particularised trust that exists in the area has not produced cooperative norms. There is a claim that bounded solidarity has dwindled following the in-migration of new and temporary residents mostly through sub-division and sub-letting of the majority of kebele-administered houses. Two development committee members of the zone also argue that those who rent makeshift beds per night carelessly destroy facilities and those who rent rooms selfishly refuse to contribute. Nonetheless, the relatively high, particularised trust in the area may have pushed aside the development of social trust as Uslaner (1999) suggests. Or the absence of trustworthy government systems and officials has prevented this particularised trust, based on proximity, from evolving into social trust as suggested by Freitag and Traunmüller (2009).

2.5 Bole Sub-City, Woreda 9

Woreda 9 of Bole sub-city is situated on the outer fringes of Addis Ababa and encompasses 769 ha of land with an estimated 85,000 residents. A report by the PPDA indicates that the sub-city took first prize in 2016 in the local government-residents' collaborative neighbourhood upgrading initiative and the contribution of Woreda 9 for this was considered large. Formal and semi-informal structures and processes for the joint initiative are identical with what exists in Woreda 6 of Addis Ketema sub-city. The woreda development council has 166 members, the woreda core development committee has 9 members, there are 9 zone committees (each with 15 members), and 555 residents have been organised into 95 one-for-five structures. Out of the nine zones in Woreda 9 of Bole sub-city, Zone 2 was selected for the research upon the recommendation of the PPDA, and sub-city and woreda heads of the PPDOs. The total number of households in the zone is 900.

Most neighbourhoods in the zone had been squatter or informal settlements on illegally purchased land from farmers. Due to the difficult terrain, irregular parcellations, and lack of social services nearby, residents had faced challenges getting the sick to hospitals. Women had to carry water uphill several times a day, and children had to travel long distances to school. Furthermore, the streets had been too narrow and muddy, making the settlements inaccessible to ambulances or fire trucks. However, except for a few, most households have recently received title deeds certifying land use rights. The survey was conducted on a randomly selected 124 households with an average family size of 3.8. The majority who participated in the survey claim that they had lived in the area since the year 2000.

Major interventions by the joint upgrading initiative in the area constituted the construction of 3.1 km of cobblestone local roads, 230 m of open ditches and 1566 m of culverts. Nearly 1.4 million birr in cash, 410,000 birr worth of labour and 87,000 birr worth of materials were mobilised from residents. Clearing right-of-way for the local roads had affected 78 residential houses in the zone. In addition, 116 fences delineating private compounds needed to be moved back and re-erected which according to the woreda PPDO office sacrificed 1,132,950 birr, making the total contribution of residents reach 3.03 million birr (€5,987 at the current exchange rate). Zone development committee members claim that while almost all had willingly agreed to either take down their walls and fences without any kind of compensation, less than half a dozen had initially refused in one mender. But most finally agreed after the committees used scare tactics by labelling those who remained adamant as “anti-development”. One claims, “A couple of residents connected to individuals in the woreda leadership are still giving us headaches”.

Even though 81% of the respondents of the survey claim that the amount of the contribution they had to make was decided by development committees,[7] 95% state that they made the contributions to the initiative willingly. And more than 65% agree that local cobblestone road construction addresses the most pressing challenge in the area. These responses indicate that the initiative had addressed the needs of the majority better than in Woreda 6 of Addis Ketema sub-city even though the level of participation and transparency in decision-making was equally questionable. Related to this, zone committee members state that they had faced relatively few challenges in collecting community contributions that were paid in three instalments. The total amount was at least 16 times more than the residents of Woreda 6 of Addis Ketema sub-city had to contribute annually for a project. Even though they badly needed the roads, some initially had refused to pay their share as a result of insecurities related to lack of tenure security. In the eyes of the law, they were squatters without legal rights when the project started in 2011. Moreover, mismanagement of funds by previous committees had made some suspicious. Unlike in Woreda 6 of Addis Ketema sub-city, zone committee members acknowledge and applaud the continuous support they had received from individuals in the woreda PPDO. They also used different sanctions, including refusing to write a support letter for the issuance of resident identification cards to the woreda administration for those who did not contribute. Residents who refused to contribute were also prohibited from making a private connection to the main water distribution line brought to the area through the initiative. The woreda administration had provided support for the enforcement of these sanctions.

Not all localities in the zone benefited from the infrastructure projects even though cash contributions had been collected from residents in all neighbourhoods of the zone. The size of the area vis-à-vis the number of households had required larger contributions, but sparse settlement patterns, leapfrogging and legal right issues had made some benefit more than others from the projects. For instance, a few are settled on land reserved as a green buffer zone by the local development plan for the river nearby. As a result, the authorisation of the project as well as the provision of title deeds were delayed in these settlements, which increased tenure insecurities and angered those who had already contributed the required amount of cash. For instance, out of the 240 households in one such settlement or mender, the construction of local cobblestone roads could not be extended to serve 60 households. This created a rift between zone committees and the households. As a result, these households

[7] The committees argue that this is not correct and that residents made the wrong assumption because they were the ones who had informed them.

discontinued their contribution for the joint initiative. Tension between residents of the different neighbourhoods also developed. But committee members argue that sub-city officials and civil servants created some conflicts that could have been avoided. In addition to delaying the provision of street designs, which negatively impacted community confidence in the local government and the zone development committee, the design the sub-city finally endorsed indicates that there may have been some irregularities involved. For example, the road designs introduced different right-of-ways in which next-door neighbours on the same street had to setback their properties for either an eight, seven or six metre wide local road.

The cobblestone local roads and open ditches are of better quality and still in good condition. Zone committee members disclosed that the sub-base material and cobblestones dumped on the site were guarded, used without delay and the residuals collected on the spot. The group discussion with zone committee members also reveals that the reason for the better performance in terms of community resource mobilisation and the better state of the physical infrastructure is directly or indirectly related to property right issues. One committee member expounds,

> The government does not recognise these kinds of settlements and will not provide physical infrastructure. It was a good opportunity that we did not dare miss. Those in the inner city have access to everything because of proximity and since the government provides these infrastructures [in legal settlements]. They may not feel they have to protect assets or contribute because they have not experienced what we had to go through and thus take things for granted.

Another one adds,

> The community is linked through iddirs and other [informal social] associations. And most of us are in the same trade and frequently interact to share information with regard to jobs. We had been here for several years struggling and are used to working together to solve our own problems. That is why we value and protect our assets. Everyone protects the cobblestone road in front of their house as if it was inside their sitting room.

These statements seem to verify the argument by Ansell and Gash (2008) that a history of cooperation makes collective action more likely. Working through the project with the local government was also construed as the first step to government recognition for the legalisation of the settlement and properties.

> We concede that lack of property rights was also a major motivating factor for dedicating our energy and resources to the initiative. We have been the first to arrive when they [local government actors] call meetings or ask residents to contribute to whatever cause they have on their agendas. But I can guarantee that our support has not dwindled after we received the title deeds.

Confidence in the zone committee is low for 32% of respondents, medium for 46%, and between very high and high for 22%. None of the respondents indicated that their confidence in the committee is very low. Confidence in the local government is also relatively better than in Woreda 6 of Addis Ketema sub-city. Indicative of this, service delivery by the local government is very high and high for 17%, medium for 42% and a lesser proportion than Woreda 6 residents (close to 41%) said that it is either low or very low. Corruption in the woreda administration is rated either very low or low for 53%, medium for 39% and between high and very high for the remaining 8%. However, 68% state that they do not leave their houses with the doors unlocked. Interviews with randomly selected individuals revealed that this does not necessarily mean that particularised trust or social capital in the area is lower than what exists in Woreda 6 of Addis Ketema sub-city. The location of the zone, sparse settlement patterns with vegetation and private compounds that reduce visibility make the area less secure. Social trust is nonetheless low as most interviewed had responded in the negative when asked if people can generally be trusted.

To conclude, a draft report prepared in 2017 for the PPDA by the Addis Ababa University on research conducted in 91 woredas of Addis Ababa indicates that 71% of respondents were willing to continue their contribution and participation in the local government-residents' joint upgrading initiative. The report also shows that income levels are not directly related to individuals' support, while property rights, type of tenure and corruption-related perceptions regarding the management of funds are. This research also shows that most respondents in both woredas want these kinds of initiatives to continue, but those in Woreda 6 of Addis Ketema are more reluctant than those in Woreda 9 of Bole to contribute their share. Even though most families in the investigated zones of Woreda 6 of Addis Ketema live inside public rental houses, residency of more than forty years and use rights that are passed down to children make tenants perceive these properties as their own, giving them a higher perception of tenure security than those in Woreda 9 of Bole. Strangely, this provides a stark contrast to the assertion by Agrawal and Ostrom (2001) on the implication of property rights on collective action (Table 2.1).

Social trust that could have provided the basic minimum level of trust needed for collective action and building trust is similarly low in both woredas. However, the type of social capital or particularised trust found in Woreda 6 of Addis Ketema sub-city has not enabled cooperation (in terms of acceptance, support and ownership). Moreover, misuse and mismanagement of public goods and communal facilities in Woreda 6 of Addis Ketema is as much a reaction against government systems as it is a lack of civic morality which the particularised type of trust in the area

Table 2.1 Survey results for confidence in the local government and semi-informal community development committees

	Woreda	Indicator	V. Low and low	Medium	V. High and high
Confidence in the local government	W6	Corruption W6	0.06	0.19	0.75
	W9	Corruption W9	0.53	0.39	0.08
	W6	Service delivery W6	0.77	0.21	0.02
	W9	Service delivery W9	0.41	0.42	0.17
Confidence in community development committees W9	W6		0.69	0.21	0.1
	W9		0.32	0.46	0.22

has not enabled. Yet with the contribution of government support, membership in such institutions also enhanced through years of working together to improve living conditions in underserviced squatter settlements and lack of tenure security have facilitated cooperation in Woreda 9 of Bole sub-city. Lack of confidence in the semi-informal committees or local government actors in the above two cases varies and is directly related to confidence in the respective local governments. However, irregularities in government systems, albeit inside higher hierarchies, have created social rifts between neighbourhoods and neighbours in Woreda 9 of Bole sub-city. Lack of transparency in selection of projects and project sites, and top-down imposition of the amount of contribution may have deterred overall willingness and satisfaction in Woreda 6 of Addis Ketema sub-city. In the case of Woreda 9 of Bole sub-city, overall satisfaction is better because the selected projects were more responsive to the needs of residents and the better quality of physical assets even though, likewise, there was a low level of transparency in the decision-making process. This is also directly related to norms in the local government. Finally, it is not social trust that created better opportunities for cooperation in Woreda 9, but rather years of toiling together for place making, moderate confidence in the local government, government support and property right issues, that played a great part.

2.6 Conclusion

The main conception of the state in collaborative/participatory planning discussions emphasises the government agency-community interaction (Ansell and Gash 2008) rather than taking a comprehensive view of the effect of the state or government systems on the system context in which collective action such as planning takes place. Where state characteristics or undemocratic government systems shape the system context, democratic planning processes, trust and cooperation are compromised.

Even though Freitag and Bühlmann (2009) state that trust among members in democratic systems tends to contribute towards the development of social trust, the research shows that the particularised form of trust and social capital that exists in the type of political/institutional environment found in Ethiopia does not provide a good foundation for cooperation or collaborative planning. Even Knack's (2002) argument that the effect of such ties, networks or membership on building trust remains within the network or among members falls short as the particularised form of trust that exists has not been able to create cooperation. In addition, the research provides evidence that cooperation is not motivated by shared goals or a we-intention, trust or some moral as Deutsch (1949), Tuomela and Miller (1985) and Daniel (1994) argue. Rather, people are willing to be part of a collective action in this kind of institutional/political environment only when they do not have other alternatives or are coerced. Agrawal and Ostrom's (2001) claim that individuals calculate their net benefit from collective action against their cost is not always the case because the research also shows that individuals base this calculation by weighing the net benefit against the benefit of participating others. And a history of cooperation works largely when it is based on sanctions as well as shared purposes. In addition, the research shows that membership in social associations (Paxton 2002) or networks (Paxton 2007) do not always guarantee the development of social trust and social capital that can be mobilised voluntarily for collective action or form the preconditions to institute collaborative planning or cooperation. Moreover, the discussion on civil society in the collaborative literature pertains to their role in articulating the interests of political communities (Healey 1999; Booher and Innes 2005). However, the research demonstrates that government systems also penetrate civil society organisations or community associations, diminishing their overall acceptance and their role in articulating public demand and bridging differences between the state and the people.

As de Satgé and Watson (2018: 8) argue, the "thin and instrumental assumptions" that planning theories make regarding the applicability of public participation or collaborative planning do not fit in with what is on the ground in other contexts, such as what is found in many parts of Africa. Yiftachel (2006) makes similar arguments by stating that practices and possibilities of planning in societies where there are fewer personal liberties, less stable property and different configurations of rationalities, among other things, require rethinking planning theorisation. The basic minimum level of trust needed for building trust and enabling cooperation, democratic institutions and culture for conducting democratic planning processes and ultimately building more participatory forms of democracy, or even the willingness to be part of a collective action may not be available in a political/institutional environment shaped by less democratic states and government systems. For instance, empirical studies have shown that social trust in African countries is low (Delhey and Newton 2004; Freitag and Bühlmann 2009). There are practical difficulties in accommodating collaborative planning methods even in countries with more liberal democratic traditions (Harris 2002). For instance, systems and environments that perpetuate individualism over social responsibilities hinder collective actions. Healey (1996) acknowledged the long way to transforming adversarial habits associated with neoliberal emphasis on competition and paternalistic government cultures. And there is evidence that collective action is being eroded faster in liberal democracies than in nations with more social democratic traditions (Putnam 1995; Delhey and Newton 2004; Stolle and Hooghe 2005; Freitag and Bühlmann 2009).

Therefore, as Healey (2003) argues, understanding socially constructed and embedded concepts and rationalities to pragmatically shape the process as per the specificity of local conditions of social structures, power relations and political possibilities highlights the need for designing planning processes within the framework of a given system's context. In a more democratic institutional/political environment, there may be a need to pay more attention to power structures born out of cultural specificities or the political economy. On the other hand, realities in the Global South necessitate focusing on the relationship between the government and the people, as the state in less democratic systems is stronger in penetrating society and shaping social norms and the world-view of actors (O'Donnell 1993). Hence, understanding the dynamics in state–society relations will clarify how certain aspects of social belief systems (for instance, trust and collectivism) are affected by government systems and become socially embedded. This will provide insight on how to organise more contextualised and accommodating planning processes or at least determine what works where.

References

Agger A, Löfgren K (2008) Democratic assessment of collaborative planning processes. Plan Theory 7(2):145–164
Agrawal A, Ostrom E (2001) Collective action, property rights, and decentralization in resource use in India and Nepal. Polit Soc 29(4):485–514
Albrechts L (2004) Strategic (spatial) planning reexamined. Environ Plann B Plann Des 31(5):743–758
Albrechts L (2013) Reframing strategic spatial planning by using a coproduction perspective. Plan Theory 12(1):46–63
Ansell C, Gash A (2008) Collaborative governance in theory and practice. Journal of Public Administration Research and Theory 18(4):543–571
Booher D, Innes J (2005) Comment. Plann Theory Pract 6(3):431–435
Brand R, Gaffikin F (2007) Collaborative planning in an uncollaborative world. Plan Theory 6(3):282–313
Bratton M, Van de Walle N (1994) Neopatrimonial regimes and political transitions in Africa. World Politics 46(04):453–489
Daniel M (1994) Reflections on the notion of cooperation. Anal Teach 14(2)
De Satgé R, Watson V (2018) Urban planning in the global south: conflicting rationalities in contested urban space. Springer
Delhey J, Newton, K (2004) Social trust: global pattern or Nordic exceptionalism? (No. SP I 2004-202). WZB Discussion Paper
Deutsch M (1949) A theory of co-operation and competition. Human Relations 2(2):129–152
Emerson K, Gerlak A (2014) Adaptation in collaborative governance regimes. Environ Manage 54(4):768–781
Fainstein S (2005) Planning theory and the city. J Plan Educ Res 25(2):121–130
Forester J (1987) Planning in the face of conflict: Negotiation and mediation strategies in local land use regulation. J Am Plann Assoc 53(3):303–314
Freitag M, Bühlmann M (2009) Crafting trust: the role of political institutions in a comparative perspective. Comp Pol Stud 42(12):1537–1566
Freitag M, Traunmüller R (2009) Spheres of trust: an empirical analysis of the foundations of particularised and generalised trust. Eur J Polit Res 48(6):782–803
Harris N (2002) Collaborative planning: from theoretical foundations to practice forms. In: Allmendinger P, Tewdwr-Jones M (eds) Planning futures: New directions in planning theory. Taylor and Francis, London, pp 21–43
Healey P (1996) The communicative turn in planning theory and its implications for spatial strategy formation. Environ Plann B Plann Des 23(2):217–234
Healey P (1998) Building institutional capacity through collaborative approaches to urban planning. Environ Plan A 30(9):1531–1546
Healey P (1999) Institutionalist analysis, communicative planning, and shaping places. J Plan Educ Res 19(2):111–121
Healey P (2003) Collaborative planning in perspective. Plan Theory 2(2):101–123
Hillier J (2002) Direct action and agonism in democratic planning practice. In: Allmendinger P, Tewdwr-Jones M (eds) Planning futures: new directions for planning theory, Routledge, London, pp 110–135

Huxley M, Yiftachel O (2000) New paradigm or old myopia? Unsettling the communicative turn in planning theory. J Plan Educ Res 19(4):333–342
Innes J, Booher D (2004) Reframing public participation: strategies for the 21st century. Plan Theory Pract 5(4):419–436
Khamis A, Kamel M, Salichs M (2006) Cooperation: concepts and general typology. In: 2006 IEEE international conference on systems, man and cybernetics, vol 2, pp 1499–1505
Knack S (2002) Social capital and the quality of government: evidence from the states. Am J Polit Sci 46(4):772–785
Newig J, Challies E, Jager NW, Kochskaemper E, Adzersen A (2018) The environmental performance of participatory and collaborative governance: a framework of causal mechanisms. Policy Stud J 46(2):269–297
O'Donnell G (1993) On the state, democratization and some conceptual problems: a Latin American view with glances at some postcommunist countries. World Dev 21(8):1355–1369
ORAAMP (2002) Addis Ababa city development plan (2002–2012). Addis Ababa
Ostrom E (1986) An agenda for the study of institutions. Public Choice 48(1):3–25
Paxton P (2002) Social capital and democracy: an interdependent relationship. Am Sociol Rev 67(2):254–277
Paxton P (2007) Association memberships and generalized trust: A multilevel model across 31 countries. Soc Forces 86(1):47–76
Powell W, DiMaggio P (1991) The new institutionalism in organizational analysis. University of Chicago Press
Putnam RD (1995) Bowling alone. J Democr 6(1):65–78
Robbins BG (2011) Neither government nor community alone: a test of state-centered models of generalized trust. Ration Soc 23(3):304–346
Rothstein B, Teorell JA (2008) What is quality of government? A theory of impartial government institutions. Governance 21(2):165–190
Skocpol T (1985) Bringing the state back in: strategies of analysis in current research. In Evans PB, Rueschemeyer D, Skocpol T (eds) Bringing the state back in. Cambridge University Press, Cambridge
Stolle D, Hooghe M (2005) Inaccurate, exceptional, one-sided or irrelevant? The debate about the alleged decline of social capital and civic engagement in Western societies. British Journal of Political Science 35(10):149–167
Tewdwr-Jones M (2002) Personal dynamics, distinctive frames and communicative planning. In: Allmendinger P, Tewdwr-Jones M (eds) Planning futures: new directions for planning theory, 2002. Routledge, London, pp 65–92
Tewdwr-Jones M, Allmendinger P (1998) Deconstructing communicative rationality: a critique of Habermasian collaborative planning. Environ Plan A 30(11):1975–1989
Tuomela R, Miller K (1985) We-intentions and social action. Analyse & Kritik 7(1):26–43
UN-Habitat (2010) The state of African cities: governance, inequalities and urban land markets. Nairobi, Kenya
Uslaner E (1999) Democracy and social capital. In: Warren M (ed) Democracy and trust. Cambridge University Press, Cambridge, pp 121–150
Yiftachel O (2006) Essay: re-engaging planning theory? Towards 'south-eastern' perspectives. Plan Theory 5(3):211–222
Yitbarek E (2008) Revisiting 'Slums', revealing responses: urban upgrading in tenant-dominated inner-city settlements in Addis Ababa, Ethiopia. Norwegian University of Science and Technology, Oslo

Lia Gabremariam Woldetsadik has a Ph.D. in the Art of Building and Town Planning from the Université Libre De Bruxelles. She has extensive practical experience working in the Ethiopian urban sector at different levels and is now a lecturer at the Ethiopian Institute of Architecture,

Building Construction and City Development of the Addis Ababa University. Her research interests include planning and governance regimes, government systems and institutions, state-society relations and collective actions.

Chapter 3
Street Vending in Downtown Rabat: In Resistance to Imported Urban Models

Wafae Bouallala

Abstract In recent years street vending has become a major feature of the public space in downtown Rabat, Morocco's capital city. Home to the Parliament and governmental institutions, downtown Rabat holds a powerful political symbolism in the collective representation of its inhabitants. Street vending is thus considered an intruder activity that must be banned from the area. However, history describes the downtown as a commercial zone where open-air markets—called *Souks*—were held regularly alongside brick-and-mortar shops before the advent of The French Protectorate in 1912, which transformed it into a "European zone" with a new "modern-formal" economy, pushing the local population to dwell in informal settlements and live from informal economy on the outskirts of the city. Through a historical analysis of the evolution of the downtown's public space use induced by the French urban laws and models, and perpetuated by the Moroccan policies after independence, this chapter argues that itinerant trade belongs to the downtown as a central function and key element of its urban dynamics that has been disrupted by alien policies. This study makes an original contribution by evaluating the impact of colonial urban policies on urban informality in the Moroccan context. Results suggest that street vending, now considered as misappropriation of space by authorities, could be considered as a form of resistance to imported planning models and that efficient urban interventions depend on an in-depth understanding of rooted local urban design.

Keywords Street Vending · Urban Informality · *Souks*

3.1 Introduction

Urban informality is the most dominant form of urbanisation globally. Whether in the field of informal housing or informal economy, numbers continue growing exponentially. By 2050, for instance, experts predict that up to three billion people on earth will live in informal urban environments (Samper et al. 2020). The informal

W. Bouallala (✉)
Institut National d'Aménagement et d'Urbanisme, Rabat, Morocco
e-mail: wafae.bouallala@gmail.com

G. Alem Gebregiorgis et al. (eds.), *Planning Cities in Africa*, The Urban Book Series,
https://doi.org/10.1007/978-3-031-06550-7_3

sector, for its part, provides more than 70% of total employment—and nearly one-third of GDP—in emerging markets and developing economies,[1] according to a recent study by the World Bank. Morocco falls right into these estimations with the informal economy representing 70% of employment and 30% of GDP (Lahlou et al. 2020).

The debate around the informal economy originated from the field of economics (Charmes 2012). The optimism of the modernisation theory led scholars in the 1950s and 1960s to believe that undeclared jobs in developing countries were part of a transitional phase and that wage-earning activities would eventually absorb the labour force. The problem was reduced to a cultural backwardness in comparison to the North. This optimism soon proved to be unfounded. Investigations showed that what was then called the traditional sector had not only persisted but had expanded to encompass new territories, as Hans Singer notes in his Kenya Report for the International Labour Organization (ILO) in 1970. Scholars started then to admit that this sector was destined to last, and began using the term "informal sector", coined by the British anthropologist Keith Hart in a study on Ghana in 1971, and adopted by the ILO in a study on Kenya in 1972.

Since then, debate around informal economy waxed and waned, but in recent decades, renewed interest has developed in the informal economy worldwide, both on scholarly and institutional levels. Today, "the informal economy is a field of study in its own right, drawing an increasing number of scholars from multiple disciplines ranging from economics, anthropology, and industrial relations to gender studies, political science, sociology, and urban planning" (Chen 2012). "There is also a flurry of high-profile policies being pursued by international agencies and Third World city governments to manage informality" (Roy 2005).

Urban studies focus on the spatial implications of informal economy with an emphasis on street vending as one of the largest and most visible sub-groups of the informal economy, with the most noticeable impact on the city. The planning literature that discusses the street vending topic is basically composed of a plethora of empirical analyses conducted in several countries in Africa (Akiyode 2017; Hlengwa 2016; Mitullah 2003), Asia (Kim and Labbé 2016; Meissonnier 2010; Bhowmik 2005) and Latin America (de Fatima Gomes and Réginensi 2007). They have diverse conceptual starting points and objects of analysis when summarising (non-exhaustively) the main topics that have been raised in the planning literature tackling informal economy, as Duminy explains in his paper elaborated for Women in Informal Employment: Globalizing and Organizing (WIEGO). He affirms that one of the most popular themes in recent scholarship is the analysis of power and resistance in urban space in reference to the conflict over public space and the confrontation between urban formal policies and the street traders' responses (Middleton 2003; Donovan 2008; Bromley and Mackie 2008).

However, the idea of resistance discussed in this chapter is not meant in the usual sense of street vendors covert and overt actions in response to authorities'

[1] https://www.worldbank.org/en/news/press-release/2021/05/11/widespread-informality-likely-to-slow-recovery-from-covid-19-in-developing-economies.

regulations and evictions (Hermawati et al. 2018), but is mentioned to highlight the way the itinerant trade spontaneously stood up to the test of time, especially in the geographic context of Morocco's capital, which called for the historic evolution of the activity. While laborious work has been conducted to track the evolution of informal settlements as a major aspect of urban informality (Samper et al. 2020), the evolution of street vending in a given area during a specific period of time is conventionally one of the hardest aspects to study due to its inconsistency and volatility over time.

Street vending is omnipresent in all Moroccan cities (HCP 2014), but the number of vendors and the impact of their activity on the public space differ from one city to another, and sometimes from one district to another within the same city. In Rabat, because of tight regulations, the downtown area has been an area inaccessible to street vendors for decades. These regulations that exist elsewhere but are stricter downtown, are due to the importance of the flow of traffic that operates there, but also to its proximity to the facilities and institutions that represent the central political power, namely: the Royal Palace, the Parliament, Ministry headquarters, Bank of Maghreb, and the Central Post Office. The downtown area, therefore, takes on a particular symbolism among users and decision-makers, to whom this space is tantamount to modernity and prestige, and where an activity such as itinerant trade, considered anarchic and illegal, as described in the media, cannot exist. Thus, the authorities intervene with successive eviction operations.

However, going back to the ancient history of the city of Rabat, its downtown and the evolution of its urban space and the practices that took place there, it emerges that itinerant trade has been a common practice in Rabat for centuries, under the name of *Souk,* the Arab word for marketplace.

Souk are at the heart of the traditional urban culture in Arab-Islamic societies (Zafirah 2018). They are one of the city's dominant features and are intimately associated to its development, its urban organisation, and its architecture (Awad 1989). *Souks* have been subject to multiple descriptive studies as a unique urban form with impactful social implications (Zafirah 2018; Hmood 2017; Mermier 2005; Awad 1989).

In the Moroccan context, the study of these marketplaces can be traced back to the French pre-colonial literature as intriguing spaces playing central roles in the daily political, economic, and social life of Morocco. (Kninah 2016).

In fact, much of the history of modern urban development in African countries is considered to be closely linked to the colonial era (Freund 2007; Njoh 1997; Mabogunje 1990; Santos 1975). Urban informality in the Global South, for instance, is believed to be one of the results of colonial policies (Kita and Okyere 2015; Santos 1975).

With the establishment of the protectorate over Morocco and the arrival of new French models of urban planning at the dawn of the twentieth century, the activity of these *souks* began to change in urban areas. Nevertheless, scholarly interest in *souks* did not wane, even after the independence, as extensive studies aimed to detail the urban and architectural organisation of *souks* (Troin 1975) in what has been described as a decisive progress in the related knowledge (Fay 1977). Other studies focused on the evolution of the traditional functions of *souks* (Ponasik 1963) and the spatial changes they have been through in modern times (Kania and Kałaska 2019).

Today we speak of street vending and *souks* as completely distinct activities. In fact, the literature does not mention a link between the *souks* and street vendors, at least in the Moroccan context. Through an analysis of the evolution of Rabat's urban space, we argue in this chapter that street vending is a mutated form of the ancestral *souks* that have been altered over time due to the succession of alien policies that overlooked this essential component of local life.

3.1.1 Study Area

The choice of Rabat to serve as a case study for this chapter was not accidental, yet it seemed to be a natural option. To elaborate a spatio-temporal analysis of the evolution of itinerant trade in the Moroccan context under the influence of imported urban policies, we looked for a city where urban transformations induced by the French Protectorate would be visible and could be tracked. Rabat was carefully chosen by the French administration to become the new capital, which meant that it was the first city to receive a special urban treatment. Indeed, the first urban policies and regulations developed theoretically in France were applied in Rabat, marking, thus, a neat and visible rupture with the existing urban reality and the start of a new urban era. Considering its status, Rabat developed a specific historic and political symbolism that led to its sacralisation in the collective imagination of its users and policy makers that shadowed its urban history before the protectorate.

Also, despite the lack of data about the evolution of Rabat's *souks* over time, there is an abundance of documentation, and especially city master plans (Basset 1989) from the beginning of the twentieth century until today that facilitate the spatio-temporal analysis we aim to explore of the links between the ancient and current itinerant trading traditions.

3.1.2 Research Materials and Method

To understand the process of Rabat's spatial evolution over time and then that of itinerant trade it was essential to overlap bibliographic, photographic, and cartographic evidence to extract targeted data and explore the underlying correlations.

Since our research deals with a known outcome, that of the proliferation of street vending in downtown Rabat, this chapter uses a retrospective analysis of the spatio-temporal changes that Rabat's *souks* have undergone before, during, and after the protectorate, to emphasise potential links between the itinerant activity both in it ancestral and current form.

3.2 *Souks* of Rabat Before the French Protectorate

3.2.1 Rabat's Urban Configuration Before the Protectorate

Before the arrival of the first French settlers in 1912, Rabat was a small and quiet city composed of two main urban centres: the Oudaïas *kasbah*,[2] a former fortified camp dating from the ninth century located at the mouth of the Bou-Regreg river, and the *medina*[3] confined in its Andalusian enclosure dating from the seventeenth century. Outside of this enclosure, a few monuments punctuated the landscape: the ruins of the Hassan Mosque, the Es-Sunna Mosque, the imperial palace built in the eighteenth century, and its gardens. These elements were delimited by a second enclosure built in the twelfth century by the Almohads. Beyond these external ramparts, rose a belvedere on the valley of Bou-Regreg, the ruins of Chellah, the old Marinid necropolis of the thirteenth century. Gardens, vineyards, and orange groves occupied the remaining space between the outer and inner walls, which surrounded the *medina* (Bennani 2012). The doors of the different enclosures were connected to those of the imperial palace, to Es-Sunna Mosque and to the Hassan tower via many paths (Chorfi 2003).

3.2.2 Rabat's Souk Before the Protectorate

Even though the term "informal economy" did not exist before 1975, historical evidence shows that the concept of self-subsistence economy existed centuries before in Morocco regardless of its denomination and juridical status evolution. "*Souks*" are one of the most prominent manifestations of this economy (Ponasik 1963). *Souk* is the Arab word for market; it refers to both the ephemeral open-air markets held on a regular basis, often weekly in the countryside, and the covered *bazaars* built within the ramparts of the *medinas*.

Troin (1975) suggests that *souks* are the result of the historical evolution of the activities of peddlers, whose movements progressively changed until they started gathering in fairgrounds either in rural or urban areas on fixed dates. Peddling itself is so old that it is hardly possible to trace it back in time. In the sixteenth century, Leon the African[4] spoke of it as an already ancient and well-established activity when describing the merchant caravans of Fez travelling from one market to another.

[2] Fortified citadel in North African cities.

[3] Ancient quarter of North African cities, usually.

[4] Joannes Leo Africanus (/ˌæfrɪˈkeɪnəs/; born al-Hasan ibn Muhammad al-Wazzan al-Fasi, Arabic: حسن ابن محمد الوزان الفاسي c. 1554) was a–c. 1494, Berber Andalusi diplomat and author who is best known for his book *Descrittione dell'Africa* (*Description of Africa*) centred on the geography of the Maghreb and Nile Valley. The book was regarded among his scholarly peers in Europe as the most authoritative treatise on the subject until the modern exploration of Africa. For this work, Leo became a household name among European geographers.

Fig. 3.1 Rabat's commercial strips and open-air *souks*

Like all imperial cities of Morocco,[5] Rabat's *medina* encompasses a large *souk* within its enclosure, which constitutes an intrinsic component of its social and economic life. Spatially, the *souk* comes in the form of commercial strips built around the main central mosque; their initial function has continued until today. This spatial configuration is common to all Moroccan *medinas* and is seen extensively in all Islamic cities of the past, with the exception of rare cases that present different urban forms linked to Islamic takeovers of classically planned cities or European colonial interventions (Correia and Taher 2015).

Besides the intramural *souk* of the *medina*, open-air market *souks* used to be held regularly outside of the *medina*'s ramparts several years before the protectorate. These markets are showcased clearly in old photographs and postcards of Rabat taken during the French Protectorate, which proves they were still functional at least during the first decade of this period. The following map tentatively aims to situate the location of the *souks* based on the aforementioned photographic evidence to achieve our research goal. While it cannot be exactly dated, the restitution roughly represents the first decade of the protectorate. This map seeks to provide a better grasp of the urban implications of the said open-air markets and serve the spatio-temporal analysis described as the methodology adopted for this chapter (Figs. 3.1 and 3.2).

[5] The imperial cities of Morocco are the four historical capital cities of Morocco: Fez, Marrakesh, Meknes, and Rabat.

Fig. 3.2 Old photographs and postal cards of Rabat's open-air market during the first decade of the French Protectorate. 1. Souk El Had 2. El Ouassaa 9. Souk Leghzal (*Source* Unknown photographers, 1. 1917, 2. 1910, 3. 1910)

The raison d'être of these open-air markets can be deduced from the analysis made by Ponasik[6] who relates in her work the changes in the functions of the Moroccan *souks* over time. Initially, this type of market was exclusive to the countryside; it filled fundamental political, economic, and social functions and obeyed a specific geographical distribution; the products sold were local and the exchange mode was barter. The gradual opening of the European market starting from the second half of the nineteenth century induced major changes to the structure of the *souks*, the products, and the movement of the merchants. An export–import dynamic flourished and completely reversed the economic structure of the countryside. The exchange became monetary and peasants became dependent on products they could not produce themselves; the main direct consequence of this new economic reality: the increase of peddlers and retailers. *Souaka* (merchants of *souks* in the local dialect) started looking for new markets in neighbouring towns and in this way assured the distribution of rural products in urban areas. This explanation correlates with the nature of the products sold at the doors of Rabat's ramparts (livestock, rural products) and shows the origin of the activity of its *souks* (Fig. 3.1). At the beginning of the twentieth century, *souks* around the country will witness new changes due to the establishment of the French Protectorate. The French administration will actually heavily rely on them to collect taxes, organise census, and control the population overall. To better manage their activities, walls were erected to surround *souks*, which encompassed built shops, dispensaries, offices of authorities, and usually a school, thus introducing urban forms into the countryside. Later on, many of them evolved into cities.

However, this was not the destiny of Rabat's open-air markets. While specific dates of their complete disappearance could not be found, it is safe to say that their life story is closely linked to the urban transformation that Rabat underwent under the French command and later under the Moroccan urban policies after independence.

[6] Diane Skelly Ponasik was a Peace Corps Volunteer in Morocco in the 1960s and has a Ph.D. in Anthropology with an emphasis on peasants of the Middle East. She spent ten years on and off in Morocco and then joined the United States Agency for International Development. During her twenty-seven-year career she was stationed in the Yemen Arab Republic, Mali, Egypt, Haiti, and Macedonia. She is now retired and lives with her husband in Washington, DC.

3.3 Urban Transformation of Rabat and Its Impact on *Souks'* Activity

3.3.1 Urban Transformation of Rabat

For Lyautey, the French army general designated to be the first Resident General to head the colonial administration, building a new modern capital that would mark the French colonial supremacy was almost a gesture of war (Bennani 2012). To achieve his urbanistic vision, he sought advice at the Social Museum in France that was militating to implement master plans for French cities.[7] It was at this time that Forestier, a French landscape architect, suggested to Lyautey to experiment on major Moroccan imperial cities with his "park system", a concept that he coined in 1906. Subsequently Prost, a French architect, was hired to transform Lyautey and Forestier's recommendations into an actual plan for the new capital Rabat. The principles were clear: separating the European zone from the indigenous zone, creating a clear zoning plan, protecting the *medina's* architecture, surrounding it with a *non aedificandi* zone for hygienic and military purposes, and articulating the city around a hierarchical and continuous network of open spaces, ranging from public gardens to peri-urban green belts.

Firstly, a traffic circulation plan was designed based on the existing paths that connected the *medina* to the existing political and cult facilities, namely the Royal Palace and the Es-Sunna Mosque. Along these paths, Prost drew wide roads dotted with new facilities emphasising the colonial power, namely the Parliament, the central bank, and postal services. The conversion of the same original tracks into structured roads facilitated the psychological and spatial implementation of the European city by investing the mental map of flow movements used by the local population (Basset 1989). In addition, a new tunnel allowed the construction of a railway station, and a national road crossed the city transforming it into a major transit point between the north and the south of the country (Fig. 3.3).

Special attention was given to landscape by elaborating a plan of open spaces that reproduced the elements of the park system concept sketched out by Forestier. Thus, as the latter had recommended, public parks, gardens, and open spaces were to be created inside and outside the Almohad enclosure before the neighbourhoods began to rise (Bennani 2012).

Then Prost drew up a zoning plan in which every part of the city was given a specific function while ensuring the complete separation between Moroccan and European zones. Besides the conservation of the *medina*, the separation between the two zones aimed to facilitate the control of the local population in case of riots and to protect the Europeans from potential epidemic risks. Thus, the *medina* was contained in its ramparts and prevented from growing naturally, to prove, in a way, the superiority of the western cultural values in terms of modern urban planning. In

---

[7] The first city master plan was completed in 1919. Morocco served as a laboratory to experiment with French theories of urbanism.

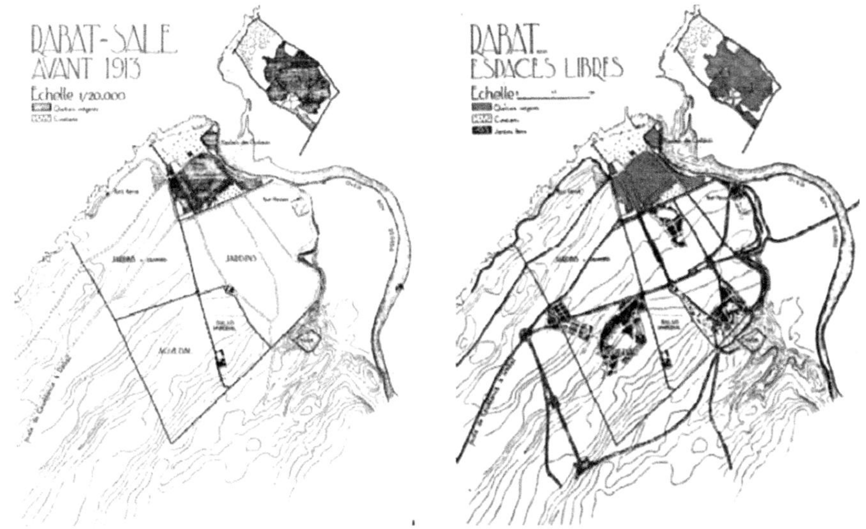

Fig. 3.3 Left: Master plan of Rabat Salé 1913 (Fonds Prost, IFA Paris). Right: Master plan of open spaces designed by Prost 1915 (*Source* L'Urbanisme aux colonies et dans les pays tropicaux, tome 1, 1932)

some references of colonial literature, the *medina* was described as tantamount to an anarchic, repetitive, and confusing space lacking logic, versus the European city that represented clarity and rationality (Mahé 1936). Except for a new small neighbourhood built for Moroccan employees, the new Rabat was completely reserved for the Europeans. In fact, Rabat's master plan was designed as if the indigenous population was stable and never going to grow (Basset 1989).

Considering this new urban configuration, the initial nucleus that formed the entire city of Rabat in previous centuries turned into the downtown area for the expanding city. Downtown Rabat was intended for residential and commercial functions and became a hub where French white-collar workers, other European blue-collar workers and Moroccans met, and the rest of the city evolved around it. Its main axis began developing in the 1920s on both sides of the route leading to the Royal Palace and the Es-sunna Mosque, on the one hand, and to the Résidence Générale (Resident Général's House) on the other. With its institutional amenities, three to five storey buildings, arcade galleries, and architectural uniqueness, downtown represented a distinguished part of the city.

3.3.2 *Socio-economic Changes Under the Protectorate: Genesis of Urban Informality*

Urban systems and cities change and develop under the impetus of various political, economic, social, and spatial factors, regardless of the geographic context. In the underdeveloped countries, exogenous interests occurring often on a global scale primarily affect the configuration of spaces and territories (Santos 1975). In Morocco, the vicissitudes that characterised its urban space and thus urban planning history were triggered by colonisation, and perpetuated by elitist policies after independence (Basset 1989).

When the French Protectorate came into force effectively in 1912, a major decision was made to transfer the capital city from Fez to Rabat. This significant move, motivated by French interests only, represents a turning point in Morocco's territorial organisation. In fact, shifting the political and economic centre of gravity from the centre of the country to its Atlantic seaboard resulted in an unprecedented rural exodus towards the coastal cities that benefited from the technological and technical progress introduced by France (Basset 1989).

These coastal cities continue to form the most important and highly active axis of the country today. At the centre of this axis, Rabat, a small and neatly limited city until the beginning of the twentieth century, on abruptly becoming the capital of the "Modern Morocco", will witness a demographic explosion and burst outside of its ramparts, pulling its neighbouring city Salé in its wake. Furthermore, the pressure generated by this sudden demographic growth will be accentuated by the segregative planning policies established by Lyautey, leading, over time, to a fragmented urban space.

Passionate about urban planning, Lyautey will rely upon it as much as he would on military mastery (Laprade 1932) and will personally engage in the creation of Rabat's novel city plan. Not altering the charm of the *medina* and building the European zone right around it was his leitmotiv to his team. Although this laudable directive saved the architectural integrity of the *medina* until today, Prost, the French architect and town planner hired to lead this mission revealed the less glorious agenda behind Lyautey's guidelines: to assure the complete separation between the European and indigenous zones for political, economic, health, and aesthetical reasons.

The Moroccan citizen then found himself confronted with a new reality where he no longer occupied the central place in the local policies. Access to goods, services, and urban space started following a new hierarchy that led to the emergence of an entire Moroccan city on the outskirts of the European city, lacking the most elementary regulations of hygiene, town planning, equipment, and construction (Dethier 1970). The exclusion from the centre of the city, the so-called European zone, drove the local population to seek places to dwell in the suburbs that were not subject to the regulations introduced by the colonial administration, thereby losing the right to their own city.

This exclusion was not by mere recklessness but was in fact the reflexion of the indirect rule "by which the 'natives' were encouraged to rule themselves along 'traditional' lines, making administration cheaper and easier (and obviating claims towards membership in the emerging colonial society) while dynamic colonial society expanded freely" as Wright (1987) cited by Freund (2007) argues.

In addition to the elitist policies, the brutal confrontation with a forced modernisation induced by entry into the sphere of the capitalist mode of production, industrialisation, and information, as Escallier (1984) analyses, and as cited by Basset (1989), squared up Rabat to two modes of spatial organisation: the top-down planning from the colonial authorities on one hand, and the bottom-up reaction from the indigenous population on the other.

Santos (1975) argues that this is a natural consequence of the modernisation in underdeveloped countries. With the technical and scientific progress and the arrival of new products, new tastes spread, old tastes subsist, and the economic apparatus faces a bipolarisation that takes place on the same territory. Moreover, modernisation means the suppression of small tasks, and creates fewer competitive jobs in a context of an increasingly active population, most of which is poorly qualified or non-qualified. This leads inevitably to the inflation of unemployment as argued by Singer (1970) and cited by Santos (1975), and the coexistence of two circuits of production, distribution, and consumption of goods and services.

In Rabat, while a new formal economy started taking hold in the centre of the city, the margins began witnessing a flourishing economy of resourcefulness: the informal economy. At the dawn of the 1920s, informality was born as a by-product of the colonial intrusion that pushed the indigenous Moroccan population to take responsibility of both housing and work away from the official unilateral urban regulations (Chouiki 2013).

3.3.3 Rabat's Souks During the Protectorate

For Rabat's *souks*, the implementation of the new urban plan and policies meant the beginning of an eventful cycle of life poorly documented. The existing literature does not specifically describe the way the open-air markets vanished, but analysis of the spatial changes induced in Rabat over time, gives informative cues about the implied consequences for these commercial structures. As discussed earlier, two types of *souks* are to be distinguished: intramural commercial strips of boutiques and open-air markets at the doors of the ramparts. While the former were well kept and preserved, the latter were subjected to various changes until they vanished, at least in their original form.

The *souks* within the *medina* have not remained totally unchanged either. The major changes linked to the new socio-economic system, the introduction of new products, the installation of new customers and the increase of traffic flow have increased the commercial dynamics inside and all around the *medina*. The function of the *souks* evolved over time and merchants had to change their commercial activity to keep up with the needs of the diverse population. New practices developed, but they remained contained within the same spatial organisation. Nevertheless, although

Fig. 3.4 Left: Souk El Ouasaa 1912, Right: Marché Central 1950- built over Souk El Ouassaa after the demolition of Bab Tben, part of the medina's rampart (*Source* Unknown photographers, Left: 1912, Right: 1950)

Prost assured the registration of all buildings and monuments with significant patrimonial value, some important inputs were induced to the old urban fabric to cope with the new commercial needs of the growing population.

Bab Tben at the end extremity of Mohammed V Street (Fig. 3.1; former Lgza Street) for example was destroyed to allow the construction of a new commercial structure: *Marché Central*. From our photographic evidence (Fig. 3.2), it is crystal clear that *Marché Central* was built on the open space called El Ouassaa where the livestock market was held daily. This market and the others described earlier in this chapter were seen as a chaotic, unsanitary, and disturbing activity by the Europeans (Kania and Kalaska 2019) (Fig. 3.4).

However, despite the obvious presence of itinerant trade in Rabat with all the urban dynamics and flows it generated, in addition to its social implications in the daily life of the indigenous population, the inclusion of this activity was completely overlooked and did not figure anywhere in Rabat's new master plan made by Prost. Policies designed in France for the European context were tested in Morocco without an effort of adaptation or participation from the locals. No spaces were designated to shelter the *souks* activity and keep them alive, nor were they converted or integrated into the new modern circuit. Even though several open spaces were designed in the downtown area, none of them was set aside for this already established commercial open-air activity. As a result, the peddlers and *souaka* of Souk el Had, Souk El Ouassaa, and Souk Leghzal were left to take responsibility for where to do business without official permission.

3.3.4 Itinerant Trade After Independence—Until Today

As a result of the coexistence of two urban planning models: the autochthon and the imported, and two urban evolution modes: the legal and illegal, the official and the parallel, Rabat's space was left chunked after the French settler left the country. After the independence, the continuity between the imported plan and the first Moroccan master plan for Rabat (*Schema Directeur*) elaborated in the beginning of the 70s was appealing. In his study of the evolution of urban planning in Rabat, Basset (1989) points out the intriguing mimicry between the two plans and questions the reasons behind it. For him, Rabat's urban crisis was met by what he called therapeutic actions to treat the crisis' symptoms instead of its root causes. Since then, the interventions on Rabat's urban space have continued to be implemented on the same principles, never considering the original urban life before the imported urban concepts.

The informal economy and its actors were once again overlooked, and informality was—and is still—considered a gangrene to be eradicated. A vision that reflected the desire to submit everything to dominant state centralism, while informality remained nothing more than a liberal, spontaneous response to a centralised and over-regulated economy (Chouiki 2013).

Downtown Rabat has kept the same functions designed by the French administration until today. In terms of commercial structures, the *medina* has sustained its commercial dynamics, while all around it new mobility hubs developed. Over the years, the pedestrian and traffic flows increased, thus attracting more street vendors (Monnet 2007). With its historic and political power symbolically built and left by the French administration, downtown was considered an untouchable part of the city and street vendors were regularly faced with forced evictions.

And yet, the successive urban interventions in the downtown area after the independence accentuated its commercial attractiveness for both formal and informal sectors. In fact, various wholesale markets and trade fairs continued to take place around the *medina* in the 80s, 90s, and early 2000s, and then were moved without offering alternatives to absorb the ever-growing itinerant activity that developed in the surroundings (Fig. 3.5).

For a decade since 2011 we have led an observational, longitudinal study of the itinerant trade evolution in downtown Rabat for research purposes: reporting changes of the spatial patterns of street vending activity allowed, documenting the trends and turning points that characterised its dynamics in this specific area over the specified time interval. Our main purpose was to identify street vendors' location

Fig. 3.5 Downtown wholesale markets in the 1970s. (1) Wholesale market (2) Bab El Had (3) Poultry market (*Source* Unknown photographers 1970)

preferences based on the spaces they would occupy the most. Being highly unstable and prone to change according to the ongoing political and economic conjuncture, it has been observed that street vending in downtown Rabat fluctuates through what seems to be perpetual cycles of prosperity, interrupted by disrupting events leading to tensions between vendors and authorities, followed by a regrouping phase and again, prosperity.

While it is interesting to study street vendors' movement during stressful times to understand resistance mechanisms, the spatial configuration during prosperity informs about the spaces that are propitious to itinerant trade when no pressure or exogenous variable is involved. The following map (Fig. 3.6) represents the specific spots in downtown Rabat that have drawn massive gatherings of street vendors during prosperous phases of the last decade.

A comparative map (Fig. 3.7) shows that itinerant trade did not disappear from the *medina* but rather flourished and expanded within its enclosure over time.

As shown on the map, today's street vendors of downtown Rabat favour spaces along the main commercial strips of the *medina*, namely Bd. Mohammed V, Souika Street and Boukroune Street, the sidewalks along the southern wall of the ramparts with significant clusters at the ramparts' portals and, interestingly, parking lots near *Marché Central* that was built over *Souk Teben-El Ouassaa* as explained earlier (Fig. 3.8) and at the *Bab El Had* portal (*Bab* being the Arabic equivalent of portal) where *Souk El Had*-Sunday Market used to be held (Fig. 3.9).

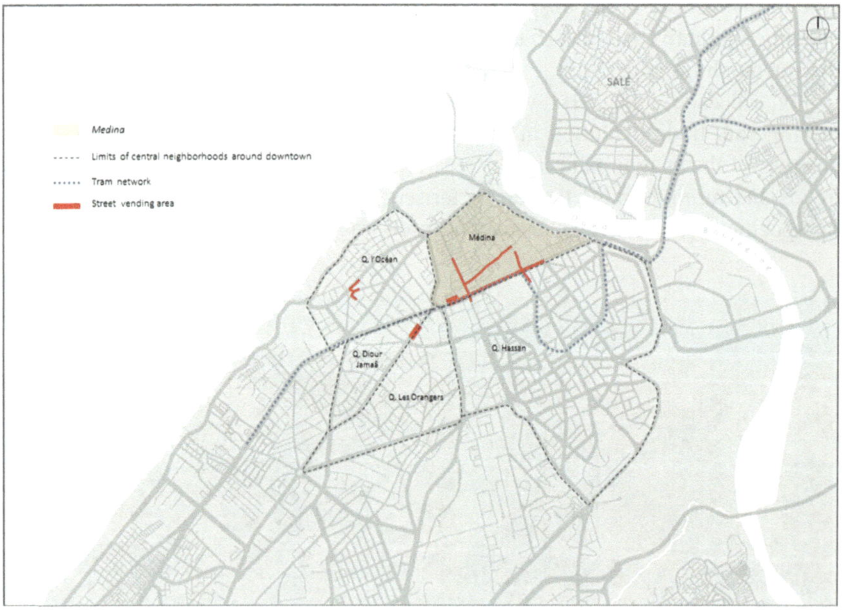

Fig. 3.6 Massive street vending areas within medina—downtown Rabat from 2011–2020

Fig. 3.7 Itinerant trade areas in the beginning of the protectorate (around 1912) and in the last decade (around 2011)

Fig. 3.8 The parking lot in front of Marché Central used by street vendors (*Source* Left: Unknown photographer 1950; Right: Taken by the author in 2011)

Fig. 3.9 Itinerant trade at the Bab El Had portal. From left to right: 1908, 1917, 2010 (*Source* Unknown photographers, 1908, 1917, 2010)

During the last decade, significant urban transformations have been conducted in downtown Rabat, primarily affecting mobility dynamics in the area. With several bus stops of major lines being created in the area, in addition to taxi ranks assuring the connection between Rabat and neighbouring cities and the implementation of the tram network, downtown Rabat became a hub where massive pedestrian and car flows met. The improved accessibility generated increased commercial value that is suitable to the prosperity of street vending (Monnet 2007). Not only the catchment area expanded, but also, more street vendors from distant neighbourhoods could easily access the area, which resulted in a rise in the number of vendors. However, and despite the visible magnitude of the activity, no official action was planned in response, other than the usual periodic evictions.

In fact, the initial commercial vocation of downtown Rabat before and during the protectorate had never been dethroned and kept increasing after independence and up until today. Street vending continued thus expanding inside and around the *medina* as shown in the following pictures (Fig. 3.10).

Today, street vending has been tremendously impacted by the pandemic and all vendors have been evicted from the downtown area to avoid gatherings. The current situation qualifies, thus, as a disturbed moment for itinerant trade in the area that opens up new questions about its uncertain future.

Fig. 3.10 Photos of different vending spots inside and around the Medina before the pandemic, 2018

3.4 Conclusion

The missing link between the original *souks* in downtown Rabat and today's street vendors lies in the history of the urban transformations Rabat underwent. When examined from a spatial point of view, regardless of the legal status, it can be concluded that itinerant trade has, in fact, never disappeared. It mutated over time and received different denominations, and, most paradoxically, turned from being a constitutive part of Rabat's downtown's history into an intrusive activity in the eyes of both policy makers and city users as observed in a former study we conducted on street vending in 2014. While peddlers were intentionally left behind and *souks* left to vanish during the protectorate, current policies seem to genuinely consider it as an intruder activity that disturbs Rabat's modernity. This calls for a reconsideration of the current orientation of planning that glorifies alien policies and laws to the detriment of the real needs of the population.

References

Akiyode AH (2017) Spaces of the informal economy: reimagining street trading through accessibility distribution analyses in Lagos. PhD Thesis. School of Geography and Planning, Cardiff University

Awad JA (1989) Islamic Souqs (Bazaars) in the urban context: the souq of Nablus. Yarmouk University, Jordan

Basset A (1989) Politiques d'urbanisme et interventions publiques dans l'urbanisation de Rabat Salé. Institut d'Aménagement Régional d'Aix-en-Provence

Bennani M (2012) Le rôle fondateur du paysage dans la création des villes coloniales marocaines: Rabat et Marrakech, deux exemples de villes-jardins. Projets de paysages

Bhowmik S (2005) Street vendors in Asia: a review. Economic and Political Weekly

Bromley RDF, Mackie PK (2008) Identifying the role of children in informal trade: evidence for urban policy. Int Dev Plan Rev 30(2):113
Charmes J (2012) The informal economy worldwide: trends and characteristics. Margin: J Appl Econ Res 6(2):103–132
Chen MA (2012) The informal economy: definitions, theories and policies. WIEGO Working Paper No. 1
Chorfi A (2003) Transformation de l'espace urbain par le Protectorat à travers le cas de la ville de Rabat. Presses universitaires François-Rabelais
Chouiki M (2013) L'informalité n'est-elle pas urbaine avant d'être économiques? Des illustrations d'une métropole du sud: Casablanca. Université Hassan II
Correia J, Taher M (2015) Traditional Islamic cities unveiled: the quest for urban design regularity. Revista Gremium
de Fatima Gomes M, Reginensi C (2007) Vendeurs ambulants à Rio de Janeiro: Street Vendors in Rio de Janeiro: civic experience and urban social practices. Revue européenne de géographie, Cybergeo
Dethier J (1970) *Soixante ans d'urbanisme au Maroc*. Bulletin économique et social du Maroc.
Donovan MG (2008) Informal cities and the contestation of public space: the case of Bogotá's street vendors, 1988–2003. Urban Studies.
Fay G (1977) J.-F. Troin, Les souks marocains : marchés ruraux et organisation de l'espace dans la moitié nord du Maroc. In: Études rurales, n° 66, 1977, pp 76–77
Freund B (2007) The African city: a history (new approaches to African history). Cambridge University Press, Cambridge
HCP (Haut-Commissariat au Plan) (2014) *Enquête Nationale sur le Secteur Informel 2013/2014*
Hermawati R, Abdoellah OS, Gunawan B, Riawanti S, Runiawati N, Paskarina C (2018) The dynamics of street vendors' resistance in Bandung City. MIMBAR: Jurnal Sosial Dan Pembangunan.
Hlengwa SP (2016) Street vending and the use of Urban Public Spaces in Tongaat Central Business District, KwaZulu Natal. PhD Thesis. University of Kwazulu-Natal
Hmood KF (2017) *Traditional markets in Islamic architecture: successful past experiences*. WIT Transactions on the Built Environment
Kania K, Kałaska M (2019) *Functional and spatial changes of souks in Morocco's imperial cities in the context of tourism development*. Miscellanea Geographica
Kim A, Labbé D (2016) Sidewalk city: remapping public space in, Ho Chi Minh City. J Soc Issues SE Asia 31(3):939–941
Kita M, Okyere SA (2015) Rethinking urban informality and informal settlements growth in urban Africa: a literature discussion. J Sust Dev Africa 17(2):101–124
Kninah A (2016) Image (s) francaise (s) Du Maroc Avant Le Protectorat: (XVIIe–XXe Siècles). Université d'Avignon et Des Pays de Vaucluse, Ecole Doctorale Culture et Patrimoine
Lahlou K, Doghmi H, Schneider F (2020) The size and development of the shadow economy in Morocco. Bank Al Maghrib
Laprade A (1932) L'urbanisme aux colonies et dans les pays tropicaux. Delagrave
Mabogunje A (1990) Urban planning and the post-colonial state in Africa: a research overview. Afr Stud Rev 33(2):121–203
Mahé Y (1936) L'extension des villes indigènes au Maroc. Université de Bordeaux, Faculté de Droit
Meissonnier J (2010) Entre ancrage et déambulation, les stratégies socio-spatiales du marchand de rue: le cas des lieux de transit à Istanbul. Commerce et mobilités, actes du colloque 2007 de la commission de géographie du commerce. Éditions Universitaires de Dijon, coll. Sociétés, 71–87
Mermier F (2005) L'urbain dans le monde musulman de Méditerranée. Institut de recherche sur le Maghreb contemporain
Middleton A (2003) Informal traders and planners in the regeneration of historic city centres: The case of Quito Ecuador. Prog Plan 59(2):71–123
Mitullah WV (2003) Street vending in African cities: a synthesis of empirical findings from Kenya, Cote D'ivoire, Ghana, Zimbabwe, Uganda and South Africa. World Bank, Washington, DC

Monnet J (2007) Ambulantage et métropolisation. European Journal of Geography, Cybergeo
Njoh AJ (1997) Colonial spatial development policies, economic instability, and urban public transportation in Cameroon. Cities 14(3):133–144
Ponasik DS (1963) Les Fonctions Modernes Du Souk Marocain
Roy A (2005) Toward an epistemology of planning. J Am Plann Assoc 71(2):147–158
Samper J, Shelby JA, Behary D (2020) The paradox of informal settlements revealed in an ATLAS of informality: findings from mapping growth in the most common yet unmapped forms of urbanization. Sustainability 12(22):9510
Santos M (1975) L'Espace Partagé: Les deux circuits de l'économie urbaine des pays sous-développés. M.Th.Génin, Paris
Troin JF (1975) Les Souks Marocains. Edisud
Zafirah MZ (2018) Souks: the heart of traditional urban life in the Middle East. MEI Kaleidoscope

Wafae Bouallala is an architect at the Moroccan Ministry of Regional and Urban Planning, Housing and City Policy, and is currently a Ph.D. student at the Institut National d'Aménagement et d'Urbanisme in Rabat, Morocco. She joined the Community Engagement, Design and Research Center (CEDaR) affiliated with the Program in Environmental Design at the University of Colorado-Boulder in the United States in December 2019 as a visiting research scholar after being awarded the Fulbright Joint Supervision Scholarship 2019–2020. Wafae's work focuses primarily on the informal economy and its spatial interaction with the city. Her current work explores the urban dynamics of informal markets in the Moroccan context combining both quantitative and qualitative methods.

Chapter 4
Revisiting Stokes' *Theory of Slums*: Towards Decolonised Housing Concepts from the Global South

Raffael Beier

Abstract Recently, large-scale housing programmes have experienced a revival in many countries of the Global South. They are criticised for their top-down, standardised, and supply-driven nature, which hardly meets people's demands. At the heart of the problem lies the concept of "material decency"—a normative and shelter-centric notion of housing, inspired by colonial planning and developmentalist thought. Many African housing programmes confuse "material decency" with the demand-driven, bottom-up concept, of adequate housing. Following this, the stigmatisation of auto-constructed neighbourhoods prevails and housing is primarily reduced to a question of material shelter. Adding to significant contributions about the need for southern perspectives on urban planning, this chapter offers an alternative entry point by revisiting Stokes' *A Theory of Slums* published in 1962. Interestingly, Stokes' theory did not deal with housing directly but focused on "slum" dwellers' socioeconomic integration and structural factors of exclusion. I argue to re-interpret Stokes' notion of barriers to social escalation as a structural discrimination of "slum" dwellers. Such stigmatisation may be read as a major reason behind the proliferation of so-called slums. Based on the author's fieldwork in Morocco and additional literature, the aim is to deconstruct the role of "material decency" and to offer pathways towards decolonised housing concepts from the Global South. For this purpose, the chapter suggests five cornerstones of adequate housing, namely subjectivity, non-materiality, flexibility, contextuality, and choice.

Keywords Decolonisation · Adequate housing · Housing theory · Informal settlements · Post-development

---

R. Beier (✉)
TU Dortmund University, Dortmund, Germany
e-mail: raffael.beier@udo.edu

G. Alem Gebregiorgis et al. (eds.), *Planning Cities in Africa*, The Urban Book Series,
https://doi.org/10.1007/978-3-031-06550-7_4

4.1 Development, Materiality, and the Stigma of "Slums"

I would like to start by retelling a brief conversation that took place in the narrow alleys of an informal settlement (*bidonville*) in one of Casablanca's working-class neighbourhoods. While I was interviewing a middle-aged housewife in front of her house, her young neighbour joined the conversation. When she heard that the housewife was feeling positive about a potential resettlement, she interrupted and said to her: "Look! Your house is out of concrete. We have wooden window frames and the house is even nicely painted in blue. What else do you want?" She responded: "I don't want to live in a *bidonville* anymore! I want to live like everybody else!".

This brief encounter highlights that informal settlements in many parts of the world represent more than just *physical* neighbourhoods with more or less precarious living conditions. These settlements exist as *social* constructs (Arabindoo 2011; Gilbert 2007; Jones 2011). They carry a stigma of otherness, backwardness, and poverty—to mention only some facets. While people that live in apartments may be seen as "developed", "modern", and "urban", informal settlement dwellers are not—even if they live in houses that share all the characteristics of *ordinary* houses (Beier 2020; Cavalcanti 2014).

By pointing out the concrete, wooden window frames, and painted walls, the young girl reminded her neighbour of the incremental developments that have transformed her neighbourhood over time. "Concrete" represents a central material feature of what Moroccans call *maisons en dur*, which literally translates as "houses built out of solid materials". As in other contexts, "concrete" relates to respected status and recognised urban citizenship (Gastrow 2017; Choplin 2020). "Window frames" are an element that other informal settlement dwellers with less developed houses frequently wished for—something distinctive that only comes at the end of an incremental construction process. Finally, she made reference to "painted walls" as an additional façade element that is an improvement to the material standards of *maisons en dur*. Referring to these basic, local status symbols, the conversation makes implicit the neighbourhood's established, basic infrastructure, such as sanitation, basic waste management, formal electricity, and running water inside the house. Notwithstanding decades of incremental development, for her neighbour instead it was clear that a *barraka*, a pejorative but frequent term for autoconstructed houses in Morocco, would never become a *maison en dur*—even if it looks the same. Decades of empty promises of resettlement, discrimination, and humiliation had given her the feeling of being a second-class citizen.

This small introductory impression emphasises an important starting point for the conceptual reflections put forward by this chapter. Terms such as *bidonville*, *favela*, or *slum* are socially constructed categories that label rather than describe certain neighbourhoods (Beier 2022; ischer 2014; Perlman 2016; Roy 2011; Valladares 2019). Likewise, they constitute the antithesis of what I suggest to call "material decency" (in Morocco, symbolised by *maisons en dur*)—a sociopolitical expectation of how "good" housing should *look* like. "Material decency" is different from a comprehensive understanding of adequate housing (although it tends to be confused with it in housing programmes), referring to a hyperbole of both technical and aesthetic standards and norms in relation to what is not (i.e. informal settlements, "slums", *bidonvilles, barraka*, etc.), rather than what it is (Beier 2021; Meth 2020). However, if society confuses adequate housing with "material decency", a denigrating label of "otherness" is likely to persist even after upgrading initiatives and decades of incremental development (Beier 2022).

Concerning Rio de Janeiro, Valladares (2019) shows impressively how after years of people-led incremental development, as well as state upgrading initiatives, still, *favelas* have remained *favelas.* Even though living conditions have significantly improved, even though brick-and-mortar houses have replaced shacks, eroding obvious differences between the formal city (the "pavement") and the *favela*, and even though they have become flourishing urban neighbourhoods that attract international tourists, the *favela* has not disappeared but persisted (Cavalcanti 2014; Fischer 2014; Jones 2011). Perlman (2005: 10) goes as far as to conclude: "The only remaining distinction between favelas [...] and the rest of the city [...] is the deeply-rooted stigma that still adheres to them". Even after resettlement away from *favelas* towards new, supposedly "formal" apartment housing, the stigma persists—continuously attached to the relocated *favelados* (Kolling 2019). Similar examples from India and Morocco show, the persistent stigma of the resettled may mingle with the marginalisation of undesired and quickly deteriorating resettlement locations—sometimes already designated as new, vertical "slums" (Beier 2019: 228f; Doshi 2019; Coelho 2016).

Early theories of Mangin (1967),Stokes (1962), and, most influentially, Turner (1968, 1969) saw the "slum" as a phenomenon that would disappear through autoconstruction and economic growth. However, even though all of this seems to have happened, at least to some extent and in some contexts, the opposite is the case—the proliferation of "slums". While some have argued that this is the result of liberal housing markets being incapable to cater for the urban poor (Berner 2016; Buckley et al. 2016; Potts 2020), others referred more to issues related to insecure and informal tenure (Payne et al. 2009; Werlin 1999; Soto 2000). However, the above-mentioned examples suggesting the persistence of "slums" even after state-led upgrading, titling, and resettlement challenge these explanations.

In this chapter, I suggest that it is an exclusionary social mindset backing the concept of "material decency" that lies behind the proliferation of "slums". It is not only because of affordability issues (Potts 2020) and housing policy failure (Berner 2016), but also because of pervasive and persistent stigmatisation of "informality" that "slums" remain. I argue that such stigma-based exclusion has its origins in colonial rule. This means the ongoing relevance of colonial spatial inequalities and building codes but even more the imposition of Western value systems and a hegemonic notion of "development" (Ziai 2013), which build the foundation of "material decency". It is not about denying unacceptable conditions in some autoconstructed neighbourhoods that require intervention according to health and safety standards to ensure adequate housing. Instead, the aim is to create awareness of an exclusionary, developmentalist mindset that prevents close engagement with a contextual, people-centred understanding of what *adequate* housing would mean. In other words, it must be challenged that an inadequately constructed, standardised apartment house at Casablanca's urban margins is implicitly associated with higher social status than a high-end, elaborately developed autoconstructed house in a central *bidonville*.

Therefore, this chapter puts forward the provocative argument that "slums" will only disappear through a profound decolonisation of dominant mindsets and a pro-active appreciation of incremental and vernacular forms of housing. Upgrading and resettlement initiatives that do not address underlying architectural and material value systems are likely to just reproduce and reinforce colonial spatial inequalities. Following post-development discourses (Ziai 2013; Escobar 2007) and calls for more Southern urban theory (McFarlane 2008; Robinson 2006; Watson 2009), this is an attempt to think of new pathways towards decolonised housing concepts.

As a starting point, the chapter re-reads Stokes' 1962 *Theory of Slums*, which although it coined the famous concepts of "slums of hope" and "slums of despair", has attracted rather limited attention compared to the influential work of Turner (Harris 2003). Unlike Turner, Stokes does not speak about any physical characteristics of "slums" but, instead, focuses on economic integration and structural barriers to social escalation. Building on Stokes' revisited theoretical argument, the chapter reviews large-scale housing programmes in different African countries, focusing on differences between proclaimed "adequate housing" and the programmes' driving concerns about "material decency". Finally, the chapter emphasises the developmentalist foundations of "material decency" and concludes by suggesting five demand-driven cornerstones for future decolonial housing concepts—namely, subjectivity, non-materiality, flexibility, contextuality, and choice.

4.2 Stokes' *Theory of Slums* Revisited

During the mid-twentieth century, social scientists discovered shantytowns and "slums" as research objects—mostly in Latin America (Valladares 2019: 65ff). During these "golden years" of developmentalism and modernisation theory, autoconstructed neighbourhoods were increasingly considered as a developmental problem, symbols of rural backwardness and urban poverty and yet a common target of evictions (Fischer 2014). Scholars like Frankenhoff (1967), Mangin (1967), Stokes (1962) and, most influential, Turner (1967, 1968, 1969), were among the first to challenge the predominant politics of eviction, positioning themselves in favour of in-situ solutions and incremental construction. While these early works shared a basic adherence to modernisation theory, believing that through enhanced and inclusive economic growth, "slums" would naturally disappear, they did not foresee this to happen in the immediate future. Turner (1969: 526) remarked, "slums" would continue to exist "as long as the poor remain poor". Likewise, Mangin (1967: 89) noted: "I heard of very few families returning to the country. The city growth and the squatter settlements are permanent developments".

With these statements, Turner and Mangin point to two common key drivers behind the growth of autoconstructed neighbourhoods: inclusive economic growth and urban population growth (Fox 2014). Both variables are crucial to Stokes' (1962)*Theory of Slums*, which argues "slum formation depends on the rate of in-migration as well as on the rate of integration or absorption of [its dwellers by the urban job market]" (1962: 191). While this central argument is certainly biased towards outdated and simplistic statements on rural-to-urban migration as the main driver of "slum" growth, it can be empirically supported if one replaces "in-migration" with "urban population growth". This acknowledges that much population growth of older precarious neighbourhoods can be explained through natural population growth as well as intra-urban migration. However, more relevant for this chapter is Stokes' (1962: 191) addition that "slum formation depends on the existence of barriers to escalation". The barriers may relate to "abilities" and to what he calls "caste", referring to structural forms of discrimination based on skin colour or religion, among others.

Unlike the architect Turner, the economist Stokes does not focus on the materiality of housing but on socioeconomic integration. Stokes' well-known distinction between slums of "hope" and "despair" is about people's very personal estimations of their chances to be absorbed by the job market and to move up the social ladder (Stokes 1962: 189). At first glance, a theory of "slums" without emphasis on material housing quality, standards, and tenure may seem strange, but, as I will show further below, it may help to counter many governments' adherence to "material decency" in housing policies. A second difference to the work of Turner is that in Stokes' theory "slums" appear to be a phenomenon equally present in both the Global North *and* the Global South. Hence, despite quite a lot of outdated and even racist language, this is a refreshing point of view towards a phenomenon, which has recently become the icon of the "undeveloped" and "chaotic" megacity of the Global South (Gilbert 2007; Huchzermeyer 2011; Roy 2011).

Indeed, Stokes' theory may appear outdated and to some extent inconsistent. For example, when he points to the "spontaneous origin" of "slums" (1962: 188), before later using Victorian row housing in Boston's South End as an example (1962: 192). Likewise, his black-and-white categorisations, as well as his strong belief in the power of economic growth and liberal market forces, are certainly a matter of contestation (Potts 2020). However, core elements of his theory still offer a good point of reflection on today's housing policies. I will highlight three main aspects: the significance of non-physical housing functions, the significance of economic integration, and discriminatory barriers to escalation. Stokes begins with an attempt at a definition of "slums" that already shows his emphasis on non-physical, societal forces.

> Slums differ from the districts in which the lowest stratum of the integrated classes live by failing to conform to the standards which this stratum has set for itself. The distinctive feature of slums is not appearance as such, then, but the relation between the slum and its inhabitants and that neighbourhood and its inhabitants which the city regards as having met minimum liveability standards. (Stokes 1962: 188)

This definition differs strikingly from current technical definitions that rather focus on physical deficiencies and legal tenure (UN-Habitat 2003). While the latter fail to explain why established autoconstructed neighbourhoods after years of incremental development and upgrading remain to be called "slums", Stokes' more constructivist approach offers a different point of view. It emphasises the role of a superordinate city society in setting standards and defining non-integration. Explicitly, this does not just refer to "objective" appearance, but to societal relationships and socially constructed difference and hierarchy. Following from this, the meaning of the term "slum" becomes immaterial, shaped by power imbalances—very similar to Valladares' (2019) notion of the invention of the *favela*.

Then, Stokes (1962: 190) goes on stressing the relevance of the city's job market for people's urban integration and social escalation. He assumes that "slum" dwellers will learn, study, and work to access employment that allows them (or their children) to move up the social ladder—a narrative that is very much present in precarious neighbourhoods (Beier 2019; Owusu et al. 2008). Perlman's (2005) long-term research on inter-generational life trajectories of *favela* residents indeed showed that many children and the majority of grandchildren of initial interviewees had left the *favela* and integrated into the city's "formal" sectors. Others had remained by choice and invested in their houses, which over the course of four decades had considerably consolidated. Thus, precarious neighbourhoods cannot be thought detached from people's aspirations to socioeconomic integration and access to urban job markets. The latter also depends on the potential availability of urban opportunities (jobs, education, etc.) that tend to be severely constrained in contexts of urbanisation without growth (Fox 2014) and unsustainably low wages, making housing largely unaffordable (Potts 2020).

Notwithstanding these general impediments for socioeconomic integration, Stokes (1962: 189) emphasises group-specific barriers. People belonging to what he calls "non-escalator classes" are structurally discriminated against on the job market and may only obtain jobs that do not permit upward mobility. Stokes (1962: 190) thinks about minority groups discriminated against because of skin colour or religion. However, referring to the introductory examples of pervasive global stigmatisation of "slum" dwellers, I suggest considering "slum" dwellers per se as structurally discriminated against. In today's cities, many dwellers of precarious neighbourhoods—whether upgraded, incrementally developed, or resettled—face discrimination based on their address (or lack thereof). An example from my own research in an established informal settlement in Casablanca highlights this. A 30-year-old female teacher shared the following experience:

> After my bachelor, I applied everywhere but I did not even receive a response. After a while, I switched the address on my identity card, using the address of my aunt living in Ain Sebaa [a middle-class neighbourhood). Afterwards, several employers invited me. Of course, it does not prove anything, but come on. (Beier 2019: 160)

Similar testimonies may be found, among others, in the work of Coelho (2012: 54) on resettled "slum" households in Chennai, India, and the work of Owusu et al. (2008: 185) on people living in Nima, Accra, and Kolling (2019: 419) on resettled *favela* residents in Salvador, Brazil. Insights from Kerala, India, show how textile factories systematically pay lower wages to women from "slums" just because of their place of residency (Devika 2016: 203). Thus, in addition to limited job opportunities and a perpetuating housing affordability gap (Potts 2020), structural discrimination on the job market limits the ability of "slum" dwellers to move up the social ladder and, hence, contributes to the persistence of "slums".

4.3 Material Decency Versus Adequate Housing

Since the turn of the millennium, scholars such as Croese et al. (2016) and Buckley et al. (2016) have witnessed a return of standardised housing programmes. Often inspired by the wish to build "slum-free" cities (Huchzermeyer 2011), they mark a disruption from international best practice of participatory in situ upgrading of precarious neighbourhoods. Examples in Africa include the standardised housing provision schemes in Angola (Croese and Pitcher 2019; Gastrow 2017), Ethiopia (Planel and Bridonneau 2017; Keller and Mukudi-Omwami 2017), Morocco (Beier 2019, 2021; Harroud 2019), Rwanda (Nikuze et al. 2019), and South Africa (Charlton 2018a, b; Meth 2020). Contrary to Stokes' *Theory of Slums*, these large-scale housing programmes opt for a shelter-centric approach of urban "integration" that tends to confuse adequate housing with the notion of "material decency".

Although they proclaim the provision of adequate housing, standardised housing programmes are little concerned with the reasons that people settle in informal settlements and their actual advantages and disadvantages for the dwellers (Croese et al. 2016). Instead, they judge and homogenise living conditions from outside while presenting standardised shelter-centric solutions as an undisputable improvement to life in "slums". In this context, Meth (2020: 159) speaks about the rhetorical construction of a hyperbole that "vastly overstates and simplifies the contrasts between formal and informal". Moreover, focusing on the "material decency" of housing ignores the significance of other dwelling functions beyond shelter that make up the concept of adequate housing—including economic, social, and emotional dwelling functions (OHCHR and UN-Habitat 2009; Satterthwaite 2020; Turner 1968). While adequate housing is demand-driven and heterogeneous, "material decency" is a normative concept largely influenced by developmentalist thought and prone to stigmatising assumptions. Like Stokes' definition of "slums", "material decency" is used to construct difference between "formal/integrated" and "informal/non-integrated" urban space. Thereby, "material decency" creates binary forms of inclusion/exclusion, fixed at the neighbourhood level, and strongly related to aesthetics, building norms, and power. In Angola, for example, questions of legality relate to the government's perception of what is aesthetically appropriate. "According to such considerations, essentially, if something looked illegal, it was illegal" (Gastrow 2017: 228). In Morocco, King Mohammed VI (MHPV 2013) sees the demolition of *bidonvilles* and the related construction of social housing as a way to ensure both dignity to poor citizens as well as aesthetic and orderly construction.

Resulting from the common confusion of "material decency" with adequate housing, objectives of standardised housing programmes go far beyond the provision of solidly built housing units. Following what Coelho (2016: 114) refers to as "shelter poverty" (a slum-centric notion of urban poverty framing it as a mere function of housing and services), large-scale housing programmes tend to assume (or rather hope) that "decent" or "formal" housing will reduce poverty and spur various subsequent forms of improvement. In the context of South Africa, state agents and policy makers intend state houses "to provide a platform for further development of the household – improving health, prosperity and education by virtue of providing a stable, safe place to live" (Charlton 2018a: 2171). According to Morocco's King Mohammed VI (2006, in MHPV 2013: 123), decent housing "should allow citizens to [...] live a peaceful and dignified life – fundamental conditions fostering the personal growth of our youth". In Ethiopia, a "code of conduct" for condominium life illustrates "authorities' drive to get new residents to adopt new modern and individual urban mindsets" (Planel and Bridonneau 2017: 37).

However, such high-flying expectations towards the provision of housing are at odds with programmes predominantly concerned with "material decency". They tend to "overstate the socioeconomic positives presumed to be gained with the receipt of formal housing" (Meth 2020: 159). In fact, there exists no theory suggesting that materially decent housing will eventually lead to urban inclusion, reduced monetary poverty, and enhanced personal development. Instead, many state-driven housing projects conflict with the socioeconomic integration of so-called beneficiaries as they relocate them further away from sources of income (Beier 2019; Culwick and Patel 2020; Harroud 2019; Kloosterboer 2019; Meth 2020; Nikuze et al. 2019). Thus, they run counter to Stokes' emphasis on economic dwelling functions and the integrative role of urban job markets.

As the provision of "formal" and materially decent housing does not guarantee enhanced economic integration, according to Stokes, neither does it ensure social upward mobility. This is not only due to impeded physical access to employment; it also relates to sustained social barriers. As indicated at the beginning of this chapter, the stigma of the "slum" may persist even after "formalisation" and the move to new housing. Again, this may relate to the notion of "material decency". However, after housing provision has taken place, it is increasingly people themselves that are blamed for non-conformity with what society defines as "material decency". People-led practices of spatial appropriation—building extensions, commercial use of residential space, or the appropriation of public space, among others—are fuelling once again stigmas around dwellers' putative incapacity to adapt to "modern" urban lifestyles. In the context of Turkey, Erman (2016: 432) found officials of a housing company that implemented a *gecekondu* resettlement "complaining about 'unruly crowds' failing to live a 'civilized' life". Likewise, Charlton (2018a: 2172f) notes,

> Perceived deviations in housing usage [...] spark reactions from within the state which reflect the concern for the appropriate conduct of households [...] rather than a concern for how the house might be deployed to improve life prospects.

Again, adequate housing is confused with an aesthetic focus on "material decency". In fact, many practices of "re-informalisation" of space mirror the inadequateness of a supply-driven provision of standardised housing units (Beier 2021; Charlton 2018a; Erman 2016). Concerning Ethiopia's new condominiums, Kloosterboer (2019: 205f) notes typical challenges for spatial appropriation:

> The organisation of space in relocation areas [...] is confronting households with a spatial change that hinders them from continuing to practice their socially and culturally constructed activities in the same way, as they had practiced before.

In addition to people's methods of spatial appropriation, a bad location, the lack of public services (especially waste management), and inadequate, cheap construction that runs counter to notions of "material decency", will lead to a persistence of stigma and various forms of discrimination. After resettlement in Casablanca, a 19-year-old student still felt uncomfortable in writing where she lives on Facebook. Another interviewee mentioned that because of the bad infrastructure, residents still get dirty shoes—an aspect typically associated with shantytown dwellers. Finally, a 22-year-old resident reported that he was rejected when applying for a job at a telecommunications agency in the city centre, as the owner felt that his domicile was too far away, and he could not rely on him should he need someone to work immediately (Beier 2019: 229).

4.4 Western Value Systems Behind Material Decency

The above section has shown that "material decency" drives Africa's mass housing programmes. The focus is not on economic integration and the reduction of social barriers to escalation, as Stokes would have recommended. Instead of looking how to improve housing situations to make them *more* adequate (i.e. through participatory in-situ upgrading), resettlement to standardised housing is merely *assumed* to stimulate poverty reduction, personal growth, and a new "urban" and "modern" mentality. Simultaneously, the programmes sustain stigmas of precarious housing as embodying the opposite of a normative "development" that is expected to emerge from materially decent housing. Such exclusionary mindsets originate from colonial understandings of urban planning and are preserved through uncritical agendas calling for "development" (Huchzermeyer 2011; Robinson 2006; Watson 2009).

Firstly, the focus on "material decency" mirrors aspirations to Western ideas of "good urbanism" that are perceived to represent "development" and "modernity". For example, Gastrow (2017: 225) argues that the aesthetic imaginations of Angola's government "reproduce a long history in African cities of the conflation of specific imaginations of urbanism, usually drawn from Euro-America, with modernity and progress". Often, colonial powers had incorporated these aesthetic imaginations in building codes and regulations, later taken over by many post-independence regimes without much change up until today (Watson 2009: 2260). In addition, urban images have gained in importance for the international competitiveness of urban centres. Visions of "development" and "modernity" circulate globally and transfer aesthetic values through images of economically prosperous cities in the North (Robinson 2006; Watson 2014). Neighbourhoods that do not conform to these aesthetic values are increasingly under threat (Huchzermeyer 2011; Ghertner 2014).

Secondly, housing policies may perceive "material decency" as a form of emancipation from colonial repression and neglect with regards to housing. Thus, overstating the benefits of "formal" housing in South Africa emerges "from the desire to distance the new developmental state from the appalling consequences of apartheid policy, acknowledging that for many, living informally was often the only way to gain an urban foothold during Apartheid" (Meth 2020: 142f). However, seeing "material decency" as an emancipatory distinction from colonially created "informality" does not question colonial norms and values of "decent housing".

Thus, driven by "material decency", Africa's housing programmes manifest Western value systems and planning norms in the name of "development". At the same time, they denigrate incremental or "informal" construction and, hence, may become blind to improvements outside the norm. Indeed, Stokes' *Theory of Slums*, as well as recent interventions by Potts (2020), stressing the role of low wages for housing affordability, urge us to look beyond material standards. Instead, "slums" are a result of limited economic integration and structural discrimination that manifests through notions of "material decency".

Taking these aspects seriously, one might come to the conclusion that some economically well integrated "slums" would have already "disappeared" if there was no structural discrimination based on a vilifying notion of "otherness". This is the case especially for established communities characterised by owner-occupation and long traditions of incremental improvement—for example, former Karyan Central in Casablanca, the Rocinha in Rio de Janeiro, and Nima in Accra. In other words, it is stigmatisation and discrimination that does not allow us to *recognise* their disappearance, because they are challenging the concept of "development" and related Western value systems behind "material decency". It is important to highlight that I do not mean *physical* disappearance—a disastrous misunderstanding behind the international slogan *Cities Without Slums* (Arabindoo 2011; Huchzermeyer 2011). Instead, it is about the sustained *social* disappearance of "slums"—something that, as mentioned above, cannot be achieved through resettlement only. To make it even clearer, I return to the example from the very beginning of this essay. Although the house from the older woman conformed to all objective features of a "decent" house—following decades of incremental improvement and economic integration—still, she did not feel accepted as a recognised urban citizen because her neighbourhood looked *different* and was known as a "shantytown".

4.5 Toward Decolonialised Concepts of Adequate Housing

Following these thoughts, it is essential to develop new, decolonialised concepts of adequate housing that challenge the primacy of "material decency". As such, these concepts must take variegated forms of "adequate housing" seriously and especially their role for economic integration and social upward mobility. Thus, in contrast to the normative top-down imposition of "material decency" that drives mass housing programmes, new housing concepts must be developed from below, starting with an analysis of housing demands (Potts 2020). This is in line with recent work of Satterthwaite (2020: 12), who claimed that "We need a better understanding of what different low-income groups need and prioritise and the choices they make, as well as how these may change over time". With reference to examples from my own fieldwork, both in a shantytown (Er-Rhamna) and at a resettlement site (Nouvelle Lahraouiyine) in Casablanca (Beier 2019), the following five cornerstones of adequate housing demands may pave the way for more demand-driven, and decolonialised housing concepts. They are subjectivity, non-materiality, flexibility, contextuality, and choice.

The first is subjectivity. It must be acknowledged that adequate housing does not mean the same thing for everyone. Borrowing the words from Turner (1968: 355), adequate housing must be understood from the "relationship between man and environment". Therefore, it is naturally a qualitative and, hence, subjective concept that cannot be understood by looking merely at material housing standards. Consequently, it is not surprising that adequate housing may be viewed differently even within the same household. For example, an older woman in Er-Rhamna shared her strong emotional attachment to her house: "I don't want to leave, I am proud of my house! My husband and I built it with our own hands". In contrast, her two young adult daughters were looking forward to potential resettlement, as they felt uncomfortable with the intense social control within the neighbourhood. Thus, it is crucial to acknowledge different housing priorities and not to judge housing conditions in a homogenising way.

Secondly, new housing concepts must stress the non-material dimensions of adequate housing and, thus, break with reductionist understandings of housing as merely physical shelter. Instead, I suggest seeing adequate housing as an inseparable whole comprised of four dimensions—material, economic, social (including political), and emotional dwelling functions. Individuals would associate personal values to each dimension, but for everyone only the togetherness of all four dimensions would make up adequate housing. While this has been long acknowledged by international declarations (OHCHR and UN-Habitat 2009), it is time to make non-material dimensions of adequate housing more explicit. In the sense of Stokes, this would especially mean aspects of socioeconomic integration. Unfortunately, as Koenig (2018: 90f) remarks with reference to West Africa, many housing programmes assume the reconstitution of jobs and urban livelihoods after intra-urban resettlement as unproblematic. Evidence suggests rather the opposite (Beier 2019; Culwick and Patel 2020; Nikuze et al. 2019). Thus, the notion of non-materiality is significant to deconstruct common beliefs behind "material decency" as the magic solution to a diverse range of social problems associated with precarious neighbourhoods (Meth 2020; Coelho 2016).

A third cornerstone might be flexibility. Across various contexts, the incremental character of autoconstruction has been widely recognised (Bredenoord and Lindert 2010; van Noorloos et al. 2020). However, unlike Latin America, Africa's housing programmes have hardly acknowledged flexibility over time. Not only do they provide fixed, one-size-fits-all solutions to which households with very heterogeneous needs must adapt, they also actively discourage people-led change (Beier 2021; Charlton 2018a; Schramm 2017). One woman in Er-Rhamna, comparing her current home with her previous rental apartment, remarked that living in a shantytown had enabled her to develop her home according to her very own personal needs. Living together with her mentally sick husband, she was happy having been able to construct a separate room as an extension to the house, which allowed her husband to live close to her without permanently sharing the same space. To oppose the stigmatisation of incremental construction, new housing concepts need to reflect on how to account for people's changing demands for adequate housing.

Fourth, adequate housing is contextual. Different geographic locations and cultural backgrounds have historically led to different, vernacular forms of housing. However, Africa's standardised housing programmes, driven by "material decency" and cost effectiveness, are rather marked by architectural homogenisation. Notwithstanding basic differences between low-density detached housing in South Africa and high-rise condominiums in Ethiopia, housing tends to be characterised by little sensitivity to local contexts. Without falling into the trap of fetishising vernacular construction (Grubbauer 2017), I would like to highlight the need to acknowledge that adequate housing demands differ according to climatic conditions, cultural and social practices. For example, in the case of Morocco, four-storey apartment houses and regular grids have disregarded cultural preferences for semi-public space such as courtyards and cul-de-sacs, which have historically served as meeting places especially for female neighbours. Likewise, in Ethiopia, standardised condominiums show little sensitivity to local practices of socialisation such as coffee ceremonies (Planel and Bridonneau 2017). In both contexts, non-contextual architectural designs and the rigid implementation of resettlement have led to social isolation (Beier 2019; Planel and Bridonneau 2017).

Finally, adequate housing is about choice. While wealthy people may choose where and how to live based on their preferences and resources (Satterthwaite 2020: 5), liberal choice is often denied to low-income people. I do not mean that low-income households simply have less choice because of fewer financial resources (Potts 2020). Their choice is often denied for political reasons and because of discrimination. It is the government that decides what is best for them. In some countries such as Rwanda and Morocco, housing programmes request people (by force) to leave the place they live and move to houses with predefined minimum standards at a certain location they are unable to influence. In other countries such as South Africa and Ethiopia, people decide to place themselves on housing waiting lists. Once selected, house assignment leaves little room for manoeuvre. However, a lack of choice does not only occur in the context of repressive housing schemes. It also builds on notions of "material decency" and related stigmatisation denying people to stay in precarious neighbourhoods out of choice. In Morocco's shantytowns, people that would have had enough resources to buy property but decided to stay faced even more stigmatisation than others. They were considered as profit-seeking spongers waiting to benefit from state housing. Again, these points highlight Stokes' structural barriers to escalation and related stigmatisation behind the social proliferation of "slums".

To conclude, I call for new housing concepts that do not struggle to explain why 40-year-old Abdallah decided to leave his newly bought, formal apartment vacant and instead preferred staying in his family home in one of Casablanca's oldest shantytowns. We need housing concepts that do not struggle to explain why Khadija is happy to wake up in order to empty the buckets inside the house each night it rains, while for others, leaking roofs are a main reason behind their wish to move out. To overcome these putative contradictions, new housing concepts must take adequate housing seriously. A stronger emphasis on the non-material dimensions of housing, for example, through Stokes' *Theory of Slums*, may offer the necessary inspiration for reflection. Likewise, the five cornerstones of adequate housing could offer a starting point to challenge "material decency" and developmentalist thought in urban planning and housing policy.

References

Arabindoo P (2011) Rhetoric of the 'Slum': rethinking urban poverty. City 15(6):636–646

Beier R (2019) From the city to the desert: analysing shantytown resettlement in Casablanca, Morocco, from residents' perspectives. Logos, Berlin

Beier R (2020) Ganz gewöhnliche Viertel: Stigma und Realitäten in Casablancas Slum Er-Rhamna. sub\urban 8(3):73–96

Beier R (2021) From visible informality to splintered informalities: reflections on the production of 'formality' in a Moroccan housing programme. Int J Urban Reg Res 45(6):930–947

Beier Raffael (2022) Ordinary neighbourhoods. Planning Theory (online first) https://doi.org/10.1177/2F14730952221076624

Berner E (2016) Housing disablement: market failures, haphazard policies and the global proliferation of slums. In: Gomez GM, Knorringa P (eds) Local governance, economic development and institutions. Palgrave Macmillan, Basingstoke, pp 98–117

Bredenoord J, van Lindert P (2010) Pro-poor housing policies: rethinking the potential of assisted self-help housing. Habitat Int 34(3):278–287

Buckley RM, Kallergis A, Wainer L (2016) Addressing the housing challenge: avoiding the Ozymandias syndrome. Environ Urban 28(1):119–138

Cavalcanti M (2014) Threshold markets: the production of real-estate value between the "Favela" and the "Pavement." In: Fischer B, McCann B, Auyero J (eds) Cities from scratch: poverty and informality in urban Latin America. Duke University Press, Durham, NC, pp 208–237

Charlton S (2018a) Confounded but complacent: accounting for how the state sees responses to its housing intervention in Johannesburg. The Journal of Development Studies 54(12):2168–2185

Charlton S (2018b) Spanning the spectrum: infrastructural experiences in South Africa's state housing programme. Int Dev Plan Rev 40(2):97–120. https://doi.org/10.3828/idpr.2018.6

Choplin A (2020) Cementing Africa: Cement flows and city-making along the West African corridor (Accra, Lomé, Cotonou, Lagos). Urban Studies 57(9):1977–1993

Coelho K (2012) The spatial reproduction of urban poverty: labour and livelihoods in a slum resettlement colony. Econ Political Weekly XLVII (47 & 48):53–63.
Coelho K (2016) Tenements, ghettos, or neighborhoods? Outcomes of slum clearance interventions in Chennai. Rev Dev Change XXI(1):111–136
Croese S, Pitcher MA (2019) Ordering power? The politics of state-led housing delivery under authoritarianism—the case of Luanda Angola. Urban Studies 56(2):401–418
Croese S, Cirolia LR, Graham N (2016) Towards Habitat III: Confronting the disjuncture between global policy and local practice on Africa's 'Challenge of Slums.' Habitat Int 53:237–242. https://doi.org/10.1016/j.habitatint.2015.11.037
Culwick C, Patel Z (2020) Building just and sustainable cities through government housing developments. Environ Urban 32(1):133–154
Devika J (2016) Aspects of socioeconomic exclusion in Kerala, India: Reflections from an urban slum. Critical Asian Studies 48(2):193–214
Doshi S (2019) Greening displacements, displacing green: environmental subjectivity, slum clearance, and the embodied political ecologies of dispossession in Mumbai. Int J Urban Reg Res 43(1):112–132
Erman T (2016) Formalization by the state, re-informalization by the people: A Gecekondu transformation housing estate as site of multiple discrepancies. Int J Urban Reg Res 40(2):425–440
Escobar A (2007) 'Post-development' as concept and social practice. In: Ziai A (ed) Exploring post-development. Routledge, London, pp 18–31
Fischer B (2014) A century in the present tense: crisis, politics, and the intellectual history of Brazil's informal cities. In: Fischer B, McCann B, Auyero J (eds) Cities from scratch: poverty and informality in urban Latin America. Duke University Press, Durham, NC, pp 9–67
Fox S (2014) The political economy of slums: theory and evidence from sub-Saharan Africa. World Dev 54:191–203
Frankenhoff CA (1967) Elements of an economic model for slums in a developing economy. Econ Dev Cult Change 19(1):27–36
Gastrow C (2017) Cement citizens: housing, demolition and political belonging in Luanda Angola. Citizsh Stud 21(2):224–239
Ghertner DA (2014) India's urban revolution: geographies of displacement beyond gentrification. Environ Plan A 46(7):1554–1571
Gilbert A (2007) The return of the slum: does language matter? Int J Urban Reg Res 31(4):697–713
Grubbauer M (2017) In search of authenticity: architectures of social engagement, modes of public recognition and the fetish of the vernacular. City 21(6):789–799
Harris R (2003) A double irony: the originality and influence of John F.C. Turner. Habitat Int 27(2):245–269
Harroud T (2019) D'une marginalite a l'autre : Droit a la ville et pratiques de readaptation des dans les nouvelles marges urbaines de Rabat. Annales De Géographie 729(730):14–37
Huchzermeyer M (2011) Cities with slums: from informal settlement eradication to a right to the city in Africa. UCT Press, Cape Town
Jones GA (2011) Slumming about. City 15(6):696–708
Keller EJ, Mukudi-Omwami E (2017) Rapid urban expansion and the challenge of pro-poor housing in Addis Ababa Ethiopia. Africa Review 9(2):173–185
Kloosterboer M (2019) The 'new' Addis Ababa: shantytown or global city? An assessment of large-scale inner-city renewal, redevelopment and displacement for the construction of a 'new' Addis Ababa. PhD thesis, University of Glasgow
Koenig D (2018) Problems endure despite policies: urban livelihoods after forced displacement. In: Cernea MM, Maldonado JK (ed) Challenging the prevailing paradigm of displacement and resettlement: Risks, impoverishment, legacies, solutions. Routledge, Abingdon, New York, pp 75–94
Kolling M (2019) Becoming favela: forced resettlement and reverse transitions of urban space in Brazil. City & Society 31(3):413–435. https://doi.org/10.1111/ciso.12237

Mangin W (1967) Latin American squatter settlements: a problem and a solution. Lat Am Res Rev 2(3):65–98
McFarlane C (2008) Urban shadows: materiality, the 'Southern City' and urban theory. Geogr Compass 2(2):340–358
Meth P (2020) 'Marginalised formalisation': an analysis of the in/formal binary through shifting policy and everyday experiences of 'poor' housing in South Africa. Int Dev Plan Rev 42(2):139–164
MHPV (Ministère de l'Habitat, de l'Urbanisme et de la Politique de la Ville) (2013) Discours et Interviews de Sa Majesté le Roi Mohammed VI: «Habitat, Urbanisme et Politique de la Ville». https://aua.ma/wp-content/uploads/2018/02/Royaume-du-Maroc.pdf. Accessed 11 Apr 2020
Nikuze A, Sliuzas R, Flacke J, van Maarseveen M (2019) Livelihood impacts of displacement and resettlement on informal households: a case study from Kigali, Rwanda. Habitat Int 86:38–47
OHCHR (Office of the United Nations High Commissioner for Human Rights), and United Nations Human Settlements Programme (UN-Habitat) (2009) The right to adequate housing: Human rights fact sheet 21. United Nations, Geneva, New York
Owusu G, Agyei-Mensah S, Lund R (2008) Slums of hope and slums of despair: mobility and livelihoods in Nima Accra. Norwegian J Geogr 62(3):180–190
Payne G, Durand-Lasserve A, Rakodi C (2009) The limits of land titling and home ownership. Environ Urban 21(2):443–462
Perlman J (2016) Rethinking precarious neighborhoods: Concepts and consequences of marginality. In: Deboulet A (ed) Rethinking precarious neighbourhoods. Agence Française de Développement, Paris, pp 39–56
Perlman J (2005) The myth of marginality revisited: the case of Favelas in Rio De Janeiro, 1969–2003. In: Hanley LM, Ruble BA, Tulchin JS (eds) Becoming global and the new poverty of cities. Woodrow Wilson International Center for Scholars, Washington, DC, pp 9–53
Planel S, Bridonneau M (2017) (Re)making politics in a new urban Ethiopia: an empirical reading of the right to the city in Addis Ababa's condominiums. Journal of Eastern African Studies 11(1):24–45
Potts D (2020) Broken cities: inside the global housing crisis. Zed Books, London
Robinson J (2006) Ordinary cities: between modernity and development. Routledge, Abingdon, New York, NY
Roy A (2011) Slumdog cities: rethinking subaltern urbanism. Int J Urban Reg Res 35(2):223–238
Satterthwaite D (2020) Getting housing back onto the development agenda: the many roles of housing and the many services it should provide its inhabitants. Environ Urban 32(1):3–18. https://doi.org/10.1177/0956247820905212
Schramm S (2017) People's room for manoeuvre in a fragmented city: state housing in Kibera, Nairobi. Transform Crit Perspect Southern Africa 93:116–141
de Soto H (2000) The mystery of capital: why capitalism triumphs in the West and fails everywhere else. Transworld, London
Stokes CJ (1962) A theory of slums. Land Econ 38(3):187–197
Turner JFC (1967) Barriers and channels for housing development in modernizing countries. J Am Inst Plann 33(3):167–181
Turner JFC (1968) Housing priorities, settlement patterns, and urban development in modernizing countries. J Am Inst Plann 34(6):354–363
Turner JFC (1969) Uncontrolled urban settlement: problems and policies. In: Breese GW (ed) The city in newly developing countries: readings on urbanism and urbanization. Prentice-Hall, Englewood Cliffs, NJ, pp 507–534
UN-Habitat (2003) the challenge of slums: Global report on human settlements 2003. Earthscan, London and Sterling, VA
Valladares L, do Prado. (2019) The invention of the Favela. University of North Carolina Press, Chapel Hill, NC

van Noorloos F, Cirolia LR, Friendly A, Jukur S, Schramm S, Steel G, Valenzuela L (2020) Incremental housing as a node for intersecting flows of city-making: rethinking the housing shortage in the global South. Environ Urban 32(1):37–54

Watson V (2009) Seeing from the South: Refocusing urban planning on the globe's central urban issues. Urban Stud 46(11):2259–2275

Watson V (2014) African urban fantasies: dreams or nightmares? Environ Urban 26(1):215–231

Werlin H (1999) The slum upgrading myth. Urban Stud 36(9):1523–1534

Ziai A (2013) The discourse of "development" and why the concept should be abandoned. Dev Pract 23(1):123–136

Dr. Raffael Beier is an urban geographer and part of the International Planning Studies research group at TU Dortmund University, where he is conducting a DFG-funded research project on post-resettlement residential mobility in housing programmes in Ethiopia, Morocco and South Africa. Previously, he worked as a postdoctoral researcher at the Centre of Urbanism and Built Environment Studies (CUBES), the University of the Witwatersrand in Johannesburg, and at the Institute of Development Research and Development Policy of Ruhr University Bochum. He holds a joint Ph.D. degree in international development studies from Erasmus University Rotterdam and Ruhr University Bochum.

Part II
The state of Planning Education and Planning Capacity

Chapter 5
In a State of Flux: Urban Planning Programmes in Asia and Africa

Geetika Anand and Nandini Dutta

Abstract The appropriateness of planning programmes in the Global South, heavily influenced by their colonial pasts and the content developed and taught in universities of the Global North, has been widely questioned. In recent years, contemporary urban challenges, as also highlightedNational Institute of Urban Affairs by the New Urban Agenda, demand that planning education step up and be a core lever of urban transformation. Grappling with legacies from the colonial past on one hand, and looking towards achieving sustainable change in future, where does planning education in post-colonial contexts currently stand? Taking seriously the intent of the programmes, this paper asks two interrelated questions of ten Master's level planning programmes across Africa and Asia: Who is the programme intended for, and to what end? What are the various forms of knowledge the programme intends to impart, and how? This comparative, qualitative review of planning programmes from across the two regions highlights the similarities and variations in how planning and its education are viewed and approached by different institutions. With the planning discipline currently in a state of flux in post-colonial contexts, this discussion presents an opportunity for learning and innovation through South-South exchanges and partnerships—a critical, yet under-explored area for collaboration when compared with existing North–South knowledge exchange partnerships.

Keywords Planning education · Curriculum review · Post-colonial contexts

5.1 Introduction

Since its origins in the early twentieth century, planning education has been under constant review aiming to establish itself as a distinct academic discipline and stay relevant with its objectives and scope. Much of the scholarship is rooted in Europe and

---

G. Anand (✉) · N. Dutta
Indian Institute for Human Settlements, Bengaluru, India
e-mail: ganand@iihs.ac.in

N. Dutta
e-mail: ndutta@iihs.ac.in

G. Alem Gebregiorgis et al. (eds.), *Planning Cities in Africa*, The Urban Book Series,
https://doi.org/10.1007/978-3-031-06550-7_5

North America where the discipline has evolved for over a century. While planning remains a relatively "young" profession (Hou 2018) in Asia and Africa, the geography of urbanisation has been shifting to these regions (UNDESA 2018) and their need for effective, trained urban planners is considered urgent and critical. Furthermore, with the adoption of the New Urban Agenda, there is a renewed interest in "territorial planning" and its potential to realise sustainable urban development (United Nations 2017). However, there remains an acute shortage of built environment professionals (planners, architects, engineers, land surveyors) in many Asian and African countries, as reported by a recent survey of Commonwealth nations (Oborn and Walters 2020).

Consequently, as these countries begin the task of reinventing and upscaling their planning education curricula, it is worthwhile to examine the prevailing situation. Sufficient evidence points to the failure of planning education in addressing situations in several Asian and African cities (UN-Habitat 2009; Watson 2011; Mahadevia and Bhatia 2018; Taşan-Kok and Oranje 2018; Denoon-Stevens et al. 2020). However, a detailed assessment of existing planning programmes across all countries is not available. Given this situation, this paper focuses on the intent of planning programmes situated in these regions. The paper asks and attempts to answer two interrelated questions by juxtaposing select planning programmes across Asia and Africa: (i) who is the programme intended for, and to what end; and (ii) what forms of knowledge does the programme intend to impart, and how?

The paper begins with an overview of the existing literature on planning education in Asia and Africa. Recognising the gap in terms of a comparative and qualitative review of planning programmes, Sect. 5.3 lays out the review framework and selection of the programmes. Findings from the selected ten programmes are presented in Sect. 5.4, followed by a discussion on emerging trends. The paper concludes with a set of observations that highlight constraints and innovations across the programmes, which we hope would encourage South-South learning and exchange as well as lay the foundation for further research and action towards rethinking and reinventing planning education in these post-colonial contexts.

5.2 The Context of Urban Planning Education in Asia and Africa

With the locus of urbanisation shifting to Asia and Africa, there is a growing scholarship on the need for and the state of planning education in these contexts. While much of the literature points towards the colonial inheritance and outdated curricula which have rendered planning education irrelevant in dealing with prevailing realities and complexities, as outlined later in this section, there are also examples of innovation and collaboration.

5.2.1 *Origins in Colonial History and Continued Influence*

Planning education in the majority of Asian and African countries is closely tied to their colonial roots, as is the development of planning as a profession. Even within the limited body of literature on planning education from these regions, much has been written on the export of planning ideas from the Global North to their southern colonies, and the power relations that shaped the way these were translated into their new contexts (UN-Habitat 2009; Watson 2011). While such foreign planning systems were originally imposed with a view to modernise, civilise and control colonial development, post-colonial governments themselves were often seen carrying forward or building on such legislation, land rights and management systems, spatial plans, administrative structures, and related frameworks.

Diaw et al. (2002) highlight that planning education in South Africa, Tanzania, and Ghana draws from colonial systems not only in terms of the actual degrees, but also the education philosophy and pedagogy. Similar concerns are voiced in relation to the Nigerian (Oduwaye and Olajide 2012) and Malawian (Blair and Manda 2016) contexts. In India, too, the orientation and content of planning programmes have originally been shaped by British planning ideologies (NIUA[1] 2017) despite a huge mismatch between the technocratic master plans developed and the realities of the largely impoverished populations to which they were applied (Mahadevia and Bhatia 2018). The establishment of the professional bodies such as the Institute of Town Planners in India and Sri Lanka were also modelled on the Royal Town Planning Institute of the UK (NIUA 2017).

In many cases, planning ideals from post-war Europe and the USA also continued to directly influence the aspirations of modern cities and planning education in Asian and African countries for decades following their independence. "Softer" forms of diffusion of planning ideas included written material and discourse, international planning consultants, international education of planners, and various decision-making bodies' engagement with development agencies (UN-Habitat 2009; Watson 2011). For many countries in these regions, early planners were trained in one or the other European nation, as in the case of Malawi (Blair and Manda 2016), and in fact, this foreign training continues to be prevalent in Asian countries (Kunzmann 2015). A number of planning programmes and their curricula, particularly in African countries, have also been shaped by funded partnerships with northern universities (Diaw et al. 2002; Odendaal 2012).

5.2.2 *Unattended Contemporary Realities*

Along with the serious concerns of colonial inheritance of planning curricula, the literature also highlights what is missing in contemporary planning education, which makes it irrelevant for the contexts it is taught in. The widening gap between theory

[1] National Institute of Urban Affairs.

and practice has been echoed by a number of Asian and African planning practitioners and educators (Todes et al. 2003; Kunzmann and Koll-Schretzenmayr 2015; Mahadevia and Bhatia 2018). Drawing from the experience of The Association of African Planning Schools (AAPS), Odendaal (2012) reports that informality, among other themes such as access to land and more recently, climate change, remains largely unaddressed in conventional planning curricula.

Speaking from the Nigerian context, Oduwaye and Olajide (2012) argue that along with a paucity of legislation, there is currently inadequate training to guide the integration of the informal sector into the urban systems of Nigeria, which are marked by high levels of informality. Denoon-Stevens et al. (2020) highlight from South Africa how land use management is seen as a key skill that is under-developed in planning education, despite being a large part of professional planning activity on the ground. They further observe that planners are required to work with "aspirations to global status at the same time as tackling the unglamorous work of attempting to mitigate grinding poverty" (Denoon-Stevens et al. 2020: 13), which has possibly led to systematic crises in the field. Meanwhile, Mahadevia and Bhatia (2018) write that planning education in India is unable to address issues of exclusion, leading to the production of even greater "informality" within Indian planning practice because the socioeconomic realities of the population being served continue to be unrecognised, unaddressed, and sometimes even actively marginalised.

Thus, not only are there missing themes and subjects, but also a disjuncture is seen between values taught in planning programmes and those prevailing at the workplace, as noted by Taşan-Kok and Oranje (2018). They highlighted frustration at the lack of guidance on the issue of power in planning curricula, leading the planners to question the utility of planning theory itself in practice. This brings planning values, and by extension, the very foundation of planning education, up for discussion. The lack of understanding of institutional and political structures and their power dynamics is a point of concern in Indian planning education as well (Chatterji and Soni 2016). As Denoon-Stevens et al. note, planning educators "expect … students to fight for abstract ideals of social justice, yet the day-to-day reality requires them to be technocrats who have to achieve this within a market-driven political system" (2020: 3).

While planning education in India has diversified over the years to reflect changing sociocultural and political realities, it is still largely focused on spatial and physical analysis; training in participatory mechanisms, for instance, is almost entirely missing from its planning curricula (NIUA 2017). Meshram and Meshram (2016) highlight that some Indian planning schools even fail to introduce students to national and state level urban development programmes, as well as the latest technological advancements in plan preparation and implementation. Traditional planning schools in India continue to propagate older physical plan-based approaches, merely adding courses involving urban development projects; instead, it is argued, today's pressing need is for greater diversification of planning education beyond the focus on producing spatial and sectoral planners (Mahadevia and Bhatia 2018).

5.2.3 Networks and Partnerships

Reflecting upon formerly colonised nations, Frank (2018) highlights that while adapting planning curricula in these contexts is necessary, it may not be straight forward; inertia and lack of resources play a large role in retaining curricula despite the obvious mismatch between planning education and urban development conditions. Since the turn of the century, however, parallel to the critiques of outdated and irrelevant planning education, there have been attempts to decolonise planning education to better reflect local circumstances and traditions, and thereby enhance the potential and relevance for the contexts in which planning is taught. Amongst the efforts, three types of networks or partnerships stand out as emerging pathways for the growth of urban planning education and its practice in Africa and Asia.

Firstly, regional and global associations have been key actors in driving forward the agenda of planning education and facilitating critical debates on its relevance, decolonisation, provision, and development (Frank and Silver 2018). One such regional network is the AAPS[2] The birth of the AAPS as a voluntary peer network in 1999 was motivated in part by the need to re-connect African planning schools in the post-apartheid and post-colonial context, and to build capacity to meet shared challenges in the region through efforts like the collaborative development of pedagogies and curricula, including a model curricular framework (Odendaal 2012; Wesely and Allen 2019). Similarly, the Commonwealth Association of Planners, a professional association, has worked to strengthen transnational connections to share and acquire knowledge, thereby, creating a more dynamic learning network throughout the North–South and the South-South (Hague 2001).

Secondly, university-university collaborations between countries, often around a joint planning programme or exchange semester, are emerging as pathways to a more multidirectional planning dialogue. Examples include partnerships between the University of Moratuwa (Sri Lanka) and LaTrobe University (Australia), and between the National University of Singapore, University of Tokyo (Japan), Tsinghua University, and Tongji University (China). These collaborative programmes have exposure and credit-sharing components, and also aid graduate entry into international job markets. Thus, alongside the efforts of decolonisation, we also see globalisation and internationalisation of the planning education and profession driving a "convergence" in planning curricula (Frank 2018: 132; Kwok 1983), possibly aided by regional differentials in funding and the position of English as a lingua franca of global planning scholarship (Stiftel and Mukhopadhyay 2007). Despite the diversity of working languages within Asia and across Africa, which remains a challenge to reciprocal knowledge-sharing and practice (Kunzmann 2015), these planning programmes provide a source of reflection and growing literature on the dynamics of new formats of international dialogue and "cross-cultural encounter" (Ratnayake and Butt 2017: 11; Wesely and Allen 2019).

[2] The AAPS has 57 member schools across 18 countries. Similarly, there is also the Asian Planning Schools Association (APSA) with 52 member schools across 14 countries (Wesley and Allen 2019).

Thirdly, multi-stakeholder and community partnerships with planning schools are highly instrumental in the deconstruction of parochial planning perspectives. These offer, among other things, opportunities to ground planning education in the contextual realities of communities and local conditions in the Global South (Watson 2009; Siame 2016). Such collaborations have lent themselves to a reframing of pedagogic instruments and practice through collaborative curriculum development, community-based planning studios, participatory research, advocacy, etc. For example, the partnership between SDI[3] and AAPS (and through it, University of Cape Town [UCT] and University of Zambia [UNZA]) is premised on the idea that in pursuit of a more inclusive urban education, "one of the most effective ways to change the mind-sets of student planners is to offer them direct experiential exposure to, and interaction with, the conditions and residents of informal settlements and slums" (Watson 2011: 23). Furthermore, the partnership approach involving multi-stakeholder engagement has been experimented with in Malawi to inculcate a system of curriculum review to embed employability within planning education (Blair and Manda 2016). It is evident that planning and its education is a growing and dynamic field in the urbanising regions of Asia and Africa, which merits attention both in terms of the breadth and the depth. As outlined in the next section, this paper adopts a qualitative framework to review and analyse planning curricula in Asia and Africa at the meta level.

5.3 Methodology and Review Framework

Using a comparative lens, the paper takes a qualitative approach to the review of planning programmes. The comparative approach to knowledge-building has been put forth in literature coming from the field of urban planning (McFarlane 2010; Robinson 2011, 2015) as well as education (Bray et al. 2014; Frank and Silver 2018). The findings of a study by Nordtveit (2016) on submissions to the Comparative Education Review indicate how the field is dominated by single case studies over comparative ones, predominantly by authors based in the area studied, and increasingly featuring quantitative studies. Within the planning education literature, few studies exist that bring together multiple planning programmes and their curricula (Friedmann 1996; van Horen et al. 2004; Ali and Doan 2006; Edwards and Bates 2011; Chatterji and Soni 2016; Prakash 2016; Sen et al. 2016; NIUA 2017; Mahadevia and Bhatia 2018; Peña 2019 and others). However, these too are either country-specific or examine planning education across countries more broadly without going into the details of specific programmes.

In this study, ten Master's level planning programmes from ten different countries in Asia and Africa are selected for review (Fig. 5.1). Many of these universities and programmes hold a significant position in advancing the agenda of planning education in their geographic contexts. Functional websites and the availability of programme documents were other practical ways of programme selection and narrowing down

---

[3] Slum Dwellers International.

Fig. 5.1 Ten postgraduate planning programmes selected for review (*Source* Authors)

the scope. The authors acknowledge that this paper does not present a complete picture of planning education in Asia and Africa; there are significant gaps in the analysis of planning education in places like mainland China, Indonesia, or Francophone Africa. However, it does provide more than a glimpse into planning education in diverse English-speaking areas of Asia and Africa, which could lead to new connections and directions in planning education research relevant to urban practice in the Global South.

Making a case for the use of different approaches in social science research, Flyvbjerg (2006) writes, large samples are useful to capture breadth while a single case study offers depth. In this paper, we position ourselves somewhere in between these two approaches, where the analysis is not based on a single case study, and the sample size has been kept small to do justice to the qualitative, in-depth nature of the review. However, we do recognise the limitation of this approach in articulating details of each and every aspect of the programmes under review and the contexts in which they are situated.

The qualitative review presented here draws its inspiration from Adamson and Morris (2014) and focuses on the "ideology" and "planned/intended" aspects of the curriculum, as articulated through publicly available information on the programmes. Thus, the paper is limited to the intent of the planning programmes; discussion on their delivery, perceptions and impacts is beyond the scope of this paper. It is hoped that this comparative analysis will lay the foundation for examining "enacted" and "experienced" aspects of curriculum (Adamson and Morris 2014) in future research. Accounting for the specificity of the planning discipline, a framework to carry out the comparative review was developed based on Davoudi's conceptualisation of planning as "practice of knowing". Davoudi (2015: 317–318) argues:

> To conceive of planning as practice of knowing requires an understanding of the complex interrelationship between knowing what (cognitive/theoretical knowledge), knowing how (skills/technical knowledge), knowing to what end (moral choices) and doing (action/practice). Together, these multiple forms of knowing provide the foundation for the art of practical judgement (wisdom). (Davoudi 2015: 317–318)

Extending this notion to higher education in planning, for planning programmes to prepare practitioners, it is important to understand to what extent these multiple, interrelated forms of knowing and knowledge are articulated and accounted for in the higher education programmes. This paper, thus, asks two interrelated, multi-faceted questions of the ten planning programmes across Asia and Africa.

- *Who is the programme intended for, and to what end?* This question looks at both programme entry and exit to examine eligibility for programme admissions as well as intended programme objectives and outcomes.
- *What are the various forms of knowledge (concepts, theories, skills, values) the programme intends to impart, and how (lectures, studio, research, training)?* Using Davoudi's conceptualisation, this line of inquiry tries to ascertain what forms of knowing (what/how/to what end/doing) find a place in these programmes, and where does the focus lie.

These framing questions are useful to not only juxtapose different programmes but also help to reveal the coherence (or lack thereof) of a programme itself. Aiming to answer these questions, the paper focuses on various aspects of programme design—vision, objectives, eligibility, core modules, electives, pedagogy, and so on, as presented in university websites, programme handbooks and brochures, and other curriculum documents. As we move into the section on findings, it is important to reiterate that this is not an exercise in comprehensive curriculum review of the education literature (Short 1991). Instead, it is an attempt to bring different planning programmes into one conversation, by examining specific questions vis-à-vis their intent, identifying key characteristics, and promoting further dialogue and research on planning education in post-colonial contexts.

5.4 A Comparison of Urban Planning Programmes Across Asia and Africa

5.4.1 Institutional Context

Table 5.1 provides details of the ten programmes selected and the universities they are a part of. With their origins in the post-independence development agenda, many of the programmes have evolved over the years in an attempt to keep abreast of changing urban conditions, new national priorities, and increased complexity of planning activity; some of these institutional efforts have been discussed by others in greater detail (Kusiima 2008; Nnkya and Lupala 2008; Oduwaye and Lawanson

Table 5.1 Planning programmes and their institutional context

S. No	Name of university	Year of origin	Timeline of planning programmes	Programme under review	Institutional affiliation	Origin of reviewed programme	Other planning programmes Offered			Accreditation/Recognition
							Undergraduate	Postgraduate	Doctoral	
1.	School of Planning and Architecture (SPA-D), New Delhi, India	1941	Started in 1955, with the establishment of first School of Town and Country Planning by the Government of India	Master of Planning (with specialisation in Urban Planning)	Department of Urban Planning	1955 Effective in 2019/2020	Bachelor of Planning	Master of Planning with specialisation in • Environ-mental Planning • Housing • Regional Planning • Transport Planning	Ph.D. in: • Urban Planning • Environ-mental Planning • Housing • Regional Planning • Transport Planning	Yes. Institute of Town Planners, India (ITPI)
2.	University of Moratuwa, Colombo, Sri Lanka	1893	Gained university status in 1972; the Dept. of Town & Country Planning was offering an M.Sc. in Town & Country Planning since 1975, until the new M.Sc. was introduced in 2012	Master of Spatial Planning, Management, & Design	Department of Town and Country Planning; Faculty of Architecture	2012–2013. Effective in 2018/2020	Bachelor of Science in Town & Country Planning	Master of Science in Environmental Planning	–	Yes. Institute of Town Planners, Sri Lanka + Planning Institute of Australia (part of La Trobe exchange programme)

(continued)

Table 5.1 (continued)

S. No	Name of university	Year of origin	Timeline of planning programmes	Programme under review	Institutional affiliation	Origin of reviewed programme	Other planning programmes Offered			Accreditation/Recognition
							Undergraduate	Postgraduate	Doctoral	
3.	National University of Singapore (NUS)	1958	Gained university status in 1980; the Faculty of Architecture first ran an MA (Urban Planning) from 1970 until 1976, when it was stopped due to lack of enrolment. Urban planning/design as a programme was not revisited until the 1990s	Master of Urban Planning	Department of Architecture; School of Design and Environment	2012 Effective in 2019/2020	–	–	Doctor of Philosophy available in Department of Architecture	None currently. In future, possibly Singapore Institute of Planners (SIP)

(continued)

Table 5.1 (continued)

S. No	Name of university	Year of origin	Timeline of planning programmes	Programme under review	Institutional affiliation	Origin of reviewed programme	Other planning programmes Offered			Accreditation/Recognition
							Undergraduate	Postgraduate	Doctoral	
4.	Hong Kong University (HKU)	1910–1912	The oldest institution of tertiary education in Hong Kong. In 1980, CUPEM/CUSUP was established for urban planning research and taught programmes (including this M.Sc.), initially as an academic unit outside the faculty structure, but brought under the Faculty of Architecture in 2008	Master of Science in Urban Planning	Department of Urban Planning and Design; Faculty of Architecture	1981 Effective in 2020, revised periodically (1, 5 years)	Bachelor of Arts in Urban Studies	• Master of Housing Management • Master of Arts in Transport Planning and Management • M.Phil.	Ph.D. programme available in the department	Yes. Hong Kong Institute of Planners (HKIP) + Royal Town Planning Institute (RTPI), UK. With specialisation, HK's Chartered Institute of Logistics & Transport

(continued)

Table 5.1 (continued)

S. No	Name of university	Year of origin	Timeline of planning programmes	Programme under review	Institutional affiliation	Origin of reviewed programme	Other planning programmes Offered			Accreditation/Recognition
							Undergraduate	Postgraduate	Doctoral	
5.	American University of Beirut (AUB), Lebanon	1866	Started at the urging of American missionaries in Lebanon and Syria, it gained its current name in 1920. The Faculty of Engineering & Architecture was established in 1951, under which the Dept. of Architecture & Design now offers its urban planning/design programmes	Master of Urban Planning and Policy	Department of Architecture and Design; Maroun Semaan Faculty of Engineering and Architecture	1998 Effective in 2019/2020	–	–	–	Not known

(continued)

Table 5.1 (continued)

S. No	Name of university	Year of origin	Timeline of planning programmes	Programme under review	Institutional affiliation	Origin of reviewed programme	Other planning programmes Offered			Accreditation/Recognition
							Undergraduate	Postgraduate	Doctoral	
6.	University of Cape Town (UCT), South Africa	1829	The Dept. of Urban and Regional Planning was founded in 1965, initially offering shorter part-time programmes, and later full-time Masters, with professional recognition from the late 1970s	Master of City and Regional Planning (linked to Honours degree in City Planning)	School of Architecture, Planning and Geomatics; Faculty of Engineering and the Built Environment	1973 Effective in 2019/2020	–	–	Ph.D. in Planning	Yes (upon completing MCRP). With 2 years supervised experience, South African Council for Planners (SACPLAN); RTPI, UK

(continued)

Table 5.1 (continued)

S. No	Name of university	Year of origin	Timeline of planning programmes	Programme under review	Institutional affiliation	Origin of reviewed programme	Other planning programmes Offered			Accreditation/Recognition
							Undergraduate	Postgraduate	Doctoral	
7.	University of Lagos (UNILAG), Nigeria	1960s	Established in 1962 to meet a national need for professional workforce; the Dept. of City & Regional Planning was formed in 1980–1981, merged briefly with the Faculty of Architecture in the late 80s, and re-emerged as an independent Dept., in 1997–1998	Master of Urban and Regional Planning	Department of City and Regional Planning; Faculty of Environmental Sciences	2003–2004 Effective in 2017–2020	B.Sc. in Urban and Regional Planning	• M.Sc. in Urban and Regional Planning M.Phil. in Urban and Regional Planning	Ph.D. in Urban and Regional Planning	Yes/soon (status unclear). Nigerian Institute of Town Planners; Town Planners Registration Council of Nigeria (TOPREC)

(continued)

Table 5.1 (continued)

S. No	Name of university	Year of origin	Timeline of planning programmes	Programme under review	Institutional affiliation	Origin of reviewed programme	Other planning programmes Offered			Accreditation/Recognition
							Undergraduate	Postgraduate	Doctoral	
8.	Makerere University, Kampala, Uganda	1922	Oldest institution in Uganda and gained independent national university status in 1970. The Dept. of Geography was established in 1940 but offered Urban planning from 1998 onwards	Master of Science in Urban Planning and Design	Department of Architecture and Physical Planning; School of Built Environment; College of Engineering, Design, Art, and Technology	1998 Effective in 2019	Bachelor of Urban and Regional Planning	Post Graduate Diploma in Urban Planning and Design	–	Not known
9.	University of Zambia (UNZA)	1965	Gained university status in 1965 but it was not until 2013 that a spatial planning programme (this M.Sc.), guided by the AAPS model curriculum, was offered to meet national planning needs following reforms in the country	Master of Science in Spatial Planning	Department of Geography and Environmental Studies; School of Natural Sciences	2013 Effective in 2019	–	–	–	Status not clear; possibly Zambian Institute of Planners

(continued)

Table 5.1 (continued)

S. No	Name of university	Year of origin	Timeline of planning programmes	Programme under review	Institutional affiliation	Origin of reviewed programme	Other planning programmes Offered			Accreditation/Recognition
							Undergraduate	Postgraduate	Doctoral	
10.	Ardhi University (ARU), Dar es Salaam, Tanzania	1956	Started as a Surveying Training School, under the Ministry of Land, Govt. of Tanzania. It trained town planners from the 1970s and underwent multiple changes over the next three decades, until it gained independent university status in 2007. Throughout, it maintained its focus on land, housing, and similar development-related concerns	Master of Science in Urban Planning and Management	Department of Urban and Regional Planning; School of Spatial Planning and Social Science	2003 Effective in 2018/19	• B.Sc. in Urban and Regional Planning • B.Sc. in Housing and Infrastruc-ture Planning • B.Sc. in Regional Develop-ment Planning	• P.G. Diploma in Urban Planning and Management • M.Sc. in Urban and Regional Development Planning and Management (jointly conducted with Technical University of Dortmund)	• Ph.D. in Urban Planning and Management by Thesis • Ph.D. in Urban Planning and Management by Coursework and Dissertation	Not known

Source Compiled by authors from university websites and Programme documents, 2020

2008; Watson 2008; Butt et al. 2013; Siame 2016; Kumar et al. 2016; NIUA 2017). A few of the programmes have been (re)launched, while others have undergone some form of curriculum revision in the early years of this decade.

The starting year of the programmes (including their predecessors) is evidence that planning as a discipline is relatively young in Asia and Africa. Within the sample, the School of Planning and Architecture, New Delhi (SPA-D), established in 1959, has the oldest postgraduate planning programme, created in 1955 when the School of Town and Country Planning was affiliated with the Delhi University. A department of Urban and Regional Planning was founded at the University of Cape Town (UCT) a decade later, in 1965 (Watson 2008). Nearly half of the programmes under review, i.e. Ardhi University (ARU), the University of Lagos (UNILAG), Hong Kong University (HKU), and the University of Moratuwa, had their origins in the 1970s and early 80 s. The urban planning programme in Makerere University was created at the turn of the century, in 1998, the same year as the programme at the American University of Beirut (AUB). The sample also includes a few recently established programmes, such as the one offered at the University of Zambia (UNZA) since 2013, and another offered at the National University of Singapore (NUS) since 2012. NUS had a part-time Master's programme in 1970, but it was discontinued in 1976 due to low enrolment. The University of Moratuwa's current planning programme was also introduced only in 2012–2013, as a joint offering between the University of Moratuwa and LaTrobe University, Australia. Three trends can be observed in terms of how these planning programmes are institutionally organised.

- Schools like SPA-D and ARU have always been set up as separate autonomous entities focused on the issues of land, surveying, architecture, and planning. Over the years, they gained the status of a university with different departments for architecture, urban planning, etc.
- Within the large universities and spanning multiple faculties and disciplines, the majority of planning programmes have close ties with architectural studies. Planning is integrated with architecture departments, as in the case of UCT, Makerere, NUS, and AUB; when there is a separate department for planning, it exists within the faculty of architecture, such as at HKU and Moratuwa.
- UNZA is the only exception where the planning programme is within the Department of Geography and Environmental Studies at the School of Natural Sciences. However, this could be a function of the current absence of architectural education at UNZA.

While several of the institutions noted in Table 5.1 offer undergraduate and multiple postgraduate planning programmes, the scope of this paper is limited to one postgraduate programme per university. It is, however, interesting to note that several universities in Africa began with undergraduate programmes before moving on to establishing graduate programmes. On the other hand, in Asian universities like SPA-D and Moratuwa, postgraduate programmes were established first, and undergraduate programmes came in about three decades later. NUS, HKU, and AUB do not have undergraduate planning degrees.

5.4.2 Programme Intent and Objectives

Planning is seen everywhere as an interdisciplinary professional degree; however, within this broader perspective, multiple similarities and differences can be observed across the programmes, in terms of how they envision planning and planners' role, who they allow into the programmes, where they see these planners going after the programme, and more.

Starting with degree titles, we see that no two programmes are named the same. Thus, to an extent, the titles themselves are suggestive of the focus of the programme. Firstly, this gives us an indication of the range in the operational scale of the planner, i.e. city, urban, regional, and more recently, spatial. Many programmes describe the planner's need to operate at or move between these multiple geographic scales. For example, NUS' Master of Urban Planning highlights the spatial planning skills that "will equip graduates with the ability to 'zoom' between scales, using the design and planning tools that are appropriate to the different scales but without losing sight of the overall picture" (NUS 2020). The idea of what constitutes the "urban scale" is understood to be a spectrum across which urban processes play out. Secondly, in some cases, these intersect, even within the titles to highlight areas of professional expertise beyond just "planning", such as policy in AUB, design in Makarere, and management in Moratuwa, ARU. This perhaps indicates the specific thrust and positioning of the programmes. For example, AUB's Master of Urban Planning and Policy positions policymaking as central to contemporary urban practice, while Moratuwa's Master of Spatial Planning, Management, and Design aims to develop planning skills but with "specific attention to urban management and design skills" (University of Moratuwa 2020), based on what they describe as a national need for such competencies.

All the programmes require applicants to hold a bachelor's degree. In terms of the discipline, HKU is the most inclusive, accepting graduates from any discipline to join their planning programme. AUB also appears to be more open, accepting students with professional or social science degrees. Most other programmes, however, require the first degree to be either in one of the spatial design or built environment professions (planning, architecture, urban design, civil engineering, surveying), or in related fields of study (geography, urban studies, development studies, etc.). SPA-D requires either a Bachelor of Planning/Architecture/Civil Engineering, or a Master of—specifically—Geography, Sociology, or Economics. Moratuwa's programme requires one to three years of work experience in the previous field of study prior to admission, as well as the first level or membership in a recognised professional institute. Makerere's programme is only open to built environment professionals, and applicants can only join after gaining two years of work experience. Overall, we see a range of disciplines across the board, but a distinct commonality skewing towards accepting those with architectural and similar spatial or built environment backgrounds.

In addition to degree titles and eligibility criteria, the programme's own stated vision and objectives, and the emphasis they are given, provide insight into how planning/planners are viewed. Two examples highlight the differences in how planning may be fundamentally understood: as a techno-managerial enterprise, in the case of SPA-D, and as a political activity, as exemplified in UCT.

The idea of problem analysis and solving is central to the programme offered at the SPA-D, which aims to create "professionals who are sensitized about the various facets of planning for human settlements and who have the required analytical skills needed for performing the assigned task related to planning and implementation" (SPA-D 2020). Emphasis within the programme is placed on various forms of "plan preparation" and implementation, as well as "new techniques such as project planning and GIS" (SPA-D 2020). In later semesters, the management aspect of cities is the core focus. This is also an imperative in Moratuwa and Makerere's programmes which aim to train their students to "conceptualize, define, and analyse design problems and opportunities at the urban scale", to "synthesize and manage strategies for implementation", and to "analyse and evaluate the performance of design projects and policies".[4] Their programmes also highlight the ability for creative "resolution of urban problems through design" (UCT 2020). However, there is also some acknowledgement that the planner must "work successfully with the public" (UCT 2020), and that there is a need for the "urban designer's public role" (UCT 2020).

The idea that planning involves social negotiation and co-creation, in addition to it comprising technical activities, is given great weight in UCT's programme statement. Here we see planning described as a "political activity" (UCT 2020) along with the idea of place-making, in reference to UN-Habitat (2009). This idea of planning as a "collective societal effort" (HKU 2020) is reflected in a few other programmes as well. For instance, strongly guided by the AAPS model curriculum, Zambia's programme also takes on this idea and promotes skills in participatory planning, the importance of ethical judgement, and engagement with issues of competing interests and differential access to power. Place-making is also a shared theme with HKU's programme, which emphasises, "Cultivating interdisciplinary, visionary and critical thinking for better place-making" and "Nurturing integrative efforts and partnership for sustainable development" (HKU 2020).

In addition to the broader role of the planner and planning, we see levels of distinct geographic focus, along with varied articulation of planning aims or challenges specific to their context, within the scope of the given programme. For example, Makerere's programme does not indicate any particular geography of practice but speaks of planning in universal terms. UNILAG's programme is proposed as a step towards developing capacity to contextually understand and gain "deeper insight" into Nigeria's planning needs (UNILAG 2017: 27). A national scope is seen in SPA-D's and Moratuwa's programmes without going into the nature of issues that arise from their contexts beyond the need for greater implementation and urban management capacities. Similarly, in AUB's programme there is a focus on "Lebanon and the region" (AUB 2018) without further context-specific details. In contrast, there are

[4] The College of Engineering, Design, Art and Technology, 2020. *Master of Science in Urban Planning and Design: Objectives of the Program.* Available from https://cedat.mak.ac.ug/graduate-programmes/master-of-science-in-urban-planning/ (Accessed 8 October 2020).

programmes with a strongly articulated thematic and geographic focus, such as those in NUS, HKU, UCT, and UNZA. In HKU's, we see them speak of "high-density environments of Hong Kong, China and Asia" and "high-rise and high-density cities" of these places (HKU 2020). They also make mention of the Asia–Pacific and the Belt and Road Regions. They aim to reshape urban planning and policy, ground professional practice, promote research, and critically review and theorise from these local contexts. Similarly, NUS' programme writes of a "rapidly urbanising world" with reference to "the experience of Singapore and cities in Asia as laboratories of planning ideas and methods, experimenting with high density living, ecological sensitivity, data science and social policies to ensure equity through development" (NUS 2020).

Half of the Asian programmes highlight the challenges of high-density settlements, in contrast to what is articulated in the African programmes. In UNZA, planning education is stated to promote "ethical, sustainable, pro-poor, gender-sensitive, and participatory planning practice", and to "ensure that the programme reflects the needs of planning in Southern Africa" (UNZA 2018). They go on to identify themes critical to African urbanisation which define this curriculum: "planning and informality; planning and climate change; planning and infrastructure; actor collaboration; and urban land markets" (UNZA 2018). UCT's programme is similarly focused on the "particular demands of cities and regions in Africa and in the Global South, in the twenty-first century. This requires us to engage with issues of poverty, inequality, informality, rapid urbanisation and environmental change" (UCT 2020). Through these four examples, we see strong local, national, and regional linkages.

All the programmes highlight multiple avenues to opportunities and placements after finishing the programme. For example, HKU's programme writes that it equips its graduates to enter "various fields of the profession, such as: urban planner and manager in the public and private sectors; professional staff in land development and management, transport, utilities companies; social and community planner in NGOs; educator and researcher in urban planning and development issues; and policy analyst and solution provider in sustainable urban development" (UNILAG 2017: 27). These are all similar to domains mentioned in other programmes, and some, like UCT, also state that graduates can "put their skills to good use in almost any part of the world" (HKU 2020). This shows the wide range of positions a planner may occupy. Interestingly, these include cases where they may find themselves in conflicting positions or contentious roles with respect to one another in practice—all under the shared umbrella of a "planner".

5.4.3 Programmes Content and Modules

All the planning programmes in the sample consist of three types of modules: classroom-based modules (including core courses and electives), field-based modules (studios/planning projects), and individual dissertation/final project. These modules are offered over two years (four semesters), except in ARU,

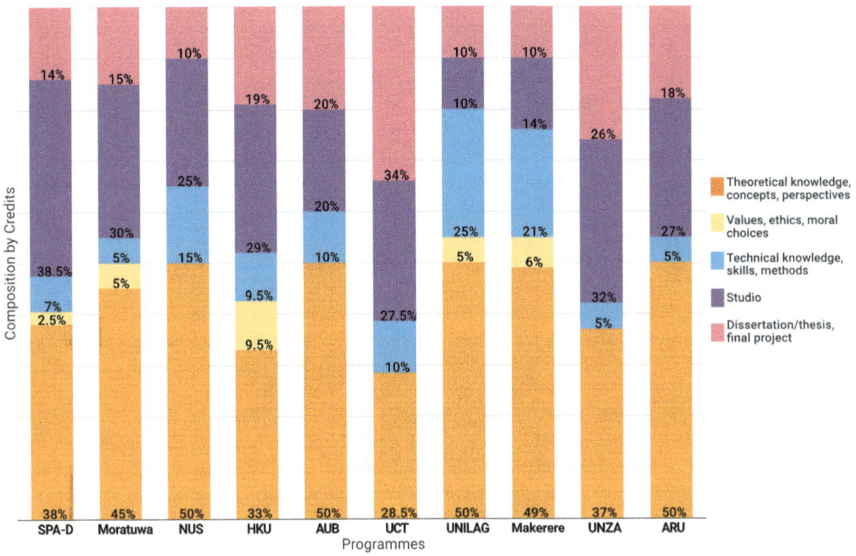

Fig. 5.2 Overall programmes' composition by types of modules based on their predominant ways of knowing (*Source* Authors 2020)

which has an 18-month-long (three semester) programme. This paper focuses on the classroom-based modules and studios. Following Davoudi (2015), and in line with the review framework set out above, in-class courses are further classified into concepts/perspectives/theoretical knowledge (knowing what); skills/methods/technical knowledge (knowing how); and values/ethics/moralchoices (knowing to what end). Guided by the principle of "learning by doing", studios and projects closely relate to the action (doing) in this framework. We recognise that these are not entirely distinct categories, and often modules speak to multiple ways of knowing, but it is useful to add this layer of analysis to understand where the focus lies and what some of the blind spots could be. Figure 5.2 presents the composition of the planning programmes, based on the overall credit weight of different types of modules. A breakdown by semester, with titles of modules, is presented in Fig. 5.3. It must be noted here that this classification is based on our interpretation and is, therefore, limited; students and faculty in these programmes, and planning educators' and practitioners' communities at large, may have a different view. Our objective here is to initiate the conversation along these lines.

5.4.3.1 Concepts/Perspectives/Theoretical Knowledge (Knowing *What*)

Most number of modules fall into the category of theoretical knowledge, with an aim to expose students to different concepts and perspectives. Upon further assessment of this category, four clusters emerge:

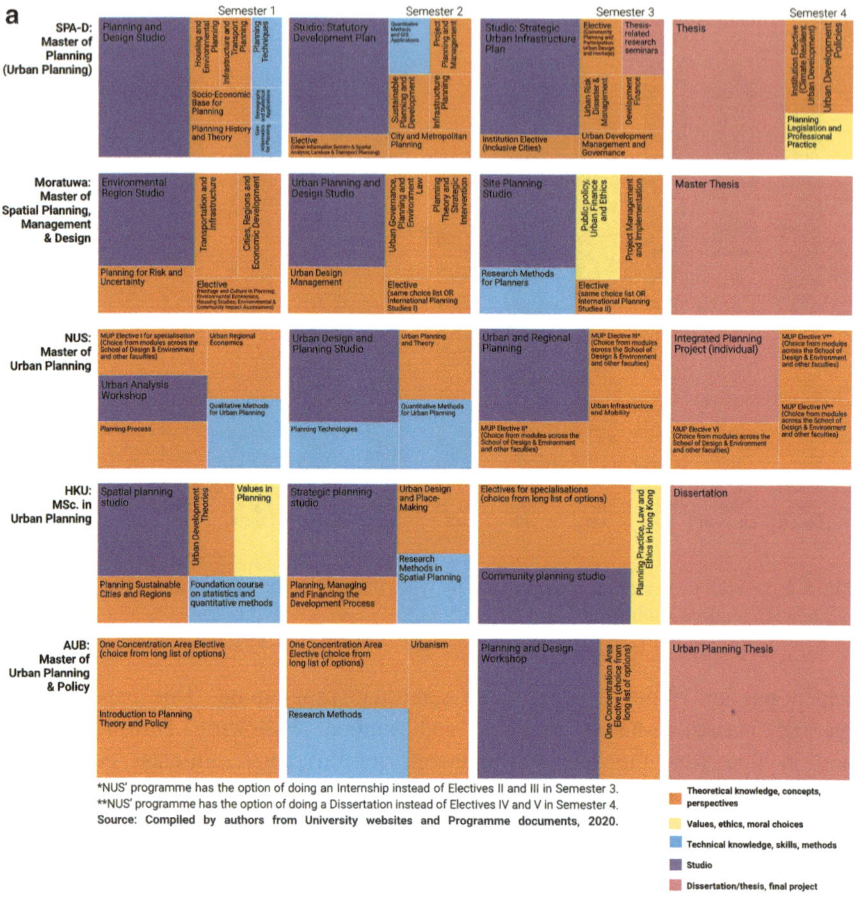

Fig. 5.3 Classification by semester of modules based on their predominant ways of knowing (*Source* Authors 2020)

- All programmes have at least one module dealing with planning theory and general planning principles.
- A second set of modules concentrate on specific sectors, with a clear focus on physical infrastructure. Seven programmes have a dedicated core module on infrastructure, four also have a specific focus on transportation either in the joint infrastructure module (Moratuwa, NUS) or as a separate module (SPA-D, UNILAG). Housing also appears in a separate core course in SPA-D and UNILAG and is included as an elective in a few others (Moratuwa, ARU).
- The third cluster includes modules with cross-cutting themes such as sustainability, governance, environment, risk, and economic development, which feature in varying degrees across many programmes. Dedicated core module(s) on planning law and governance find a place in half of the programmes. Some unique modules that merit further investigation include land-focused modules

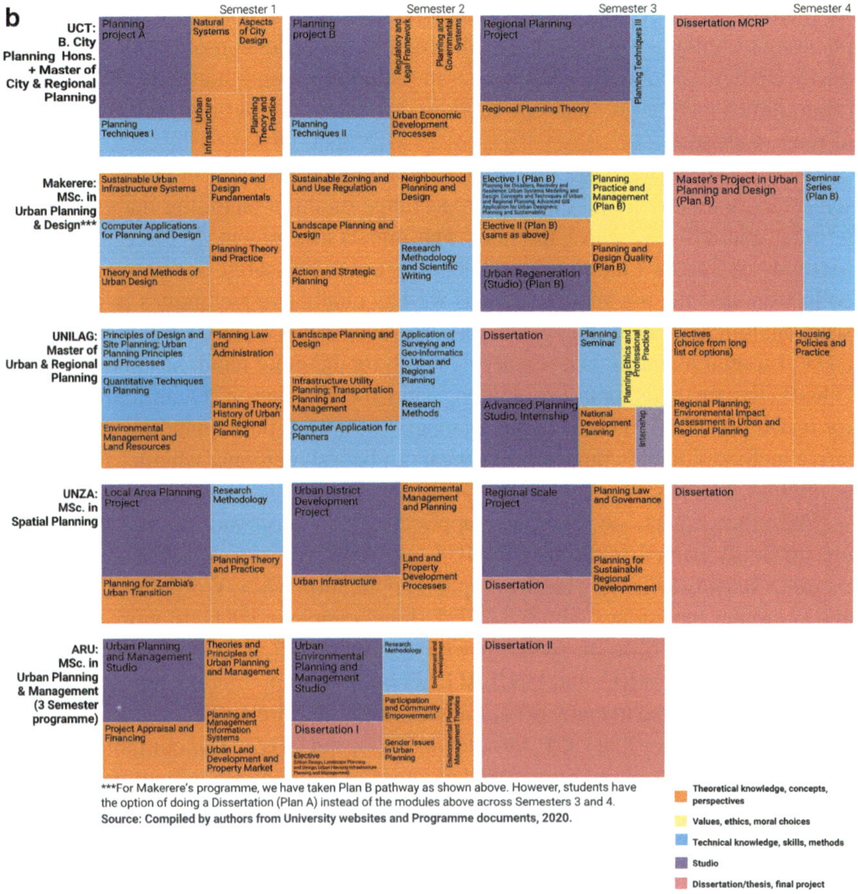

Fig. 5.3 (continued)

(ARU, UNZA); gender issues in urban planning (ARU); and inclusive cities (an elective in SPA-D). This cluster appears to be the most dynamic across the programmes, where many universities are trying to introduce modules that speak to contemporary urban challenges in cities of the Global South.

- Lastly, modules pertaining to planning instruments like project management, budgeting, and zoning, transcend the theoretical knowledge framework and have an overlap with technical knowledge.

5.4.3.2 Skills/Methods/Technical Knowledge (*Knowing How*)

There appear to be three technical aspects covered across all programmes, to varying degrees through these module(s): (1) quantitative and qualitative methods for urban planning and research; (2) computer applications and software skills (e.g. ArchGIS) for planning practice; and (3) technical report writing. A module on methods and/or planning techniques can be found in all programmes under consideration. Moratuwa, ARU, UNZA, and AUB—each have a module on research methods. The other six programmes have more than one module dedicated to building the technical skills of their students.

UCT, for example, has a module on "planning techniques" in three of their four semesters, which together cover a broad range of skills including mapping, report writing, qualitative research methods, and impact assessment. In HKU's "Research Methods in Spatial Planning" module, they cover "research design methods, data collection, and the use of statistical as well as qualitative techniques in data analysis… analytical models and evaluation and management methods that are commonly used in spatial planning and research". NUS, on the other hand, has one module on "Qualitative methods for urban planning" and a second one on "Quantitative methods for urban planning". This idea that planners must have both quantitative and qualitative methodological competency appears to be a shared one, whether for research or applied analyses.

The other aspect, i.e. computer applications and software skills for planning, are explicitly named in all but the NUS and AUB programmes. We see dedicated modules starting from very basic computer applications (MS, Word, Excel) in SPA-D and computer techniques within a foundational course in HKU, to more specialised technology-based skills in working with geo-informatics, remote sensing, etc., in SPA-D, Makerere, ARU, and UNILAG. Technology supported planning techniques are being highlighted in many programmes as a key requirement for planning practice today. In other places like UCT, Moratuwa, HKU, and UNZA, this computer/software training is located within larger mixed modules of planning techniques and methods, or is integrated with studios. While technical writing is emphasised in planning techniques modules, other forms of communication receive less attention. They are mostly subsumed within studio projects or found, if at all, as a small component of other modules.

5.4.3.3 Values/Ethics/Moral Choices (Knowing *to What End*)

Underlying values of planning programmes are visible in multiple places—vision statements, choice of themes and theoretical modules, and also where and how studios are conducted. In terms of the programme design and modules, however, limited space exists in most of the programmes to deliberate upon values and moral choices that planners are faced with. HKU has the only programme with a dedicated module on values in planning, in addition to one on professional practice, which is also included in SPA-D, Moratuwa, Makarere, and UNILAG, where issues of ethics and code of conduct are covered. In other programmes (UCT, UNZA), planning ethics are included within the larger module on planning theory and practice.

5.4.3.4 Studios and Projects (*Doing*)

Studios form the backbone of all planning programmes. They make up close to a third of overall programme credits in the majority of programmes (SPA-D, HKU, NUS, Moratuwa, UCT, ARU, UNZA), and constitute an even more critical component in each of the first two or three semesters, until the dissertation component in the final semester. AUB, Makerere, and UNILAG each have only one studio module in the entire programme and relatively smaller studio components as a percentage of the total programme credits. UNILAG's focus is on the design of new towns, capital cities, and master plans, whereas Makerere is focused on "urban regeneration" and "practical debates" and solutions. AUB, within a single studio, goes through the process of planning—from documenting, analysing, and problem framing, to conceptualising and designing interventions. In NUS, this planning process is dealt with across scales, from city area to urban design to regional planning.

The scalar progression—from local to regional—appears to be typical of many of the programmes. For seven out of the ten programmes, there is a focus on "local" in the first semester and a focus on the intermediary scale (like urban district, city, metropolitan area, etc.) in the second. UNZA, for example, moves from working at local area planning in the first semester to urban district planning in the second, followed by regional scale in the third. Moratuwa, however, is an outlier in that it has an opposite scalar progression, starting from the environmental region and moving to urban planning and design, then to site planning in the third semester. Overall, the third semester is where we see the most thematic variations in the studio across the programmes.

When comparing studios at SPA-D to those at HKU, different approaches to studio settings come to light. SPA-D is structured more in terms of statutory plans and planning products, rather than explicitly around scale, though there is an aspect of the typical scale progression seen here as well. In the first semester, it is framed around an area appreciation study and City Development Plan, followed by a Tier II City-level statutory development plan in the second semester, and finally an urban infrastructure plan, feasibility study and its Detailed Project Report in the third. HKU, on the other hand, is structured primarily around three types of planning/planners: spatial planning in the first semester, strategic planning in the second, and community planning in the third. While all the studios are set in urban areas within their contexts, some of the programmes (e.g. UNZA, UCT) actively collaborate with partners, for example, Slum Dwellers International (SDI), to make the experience even more grounded.

5.5 Discussion and Conclusion

The first thing that can be concluded from this review is the diversity found among urban planning programmes at different institutions. We see planning presented in varied manifestations, and indications that multiple influences over time have shaped each programme's formation and greater understanding of planning/planners.

While the evidence does confirm the rootedness of several of these programmes in their colonial pasts, with relics from that era of planning still present in their design and modules, there are signs to indicate a shift from this origin in all programmes to varying degrees. In some, the need for this shift appears more articulated and has perhaps been given greater impetus through active and critical regional debates and dialogues in recent decades, as observed in the case of UCT or UNZA through AAPS, while in others, such as SPA-D or UNILAG, efforts to reframe planning education appear as yet to be in early stages. While the intention and the energy with which the change has taken place may differ, it is clear there has been some evolution in the planning curricula in Asia and Africa from colonial and even post-colonial era development logics.

Some of these changes echo contemporary national as well as shared bilateral or international development agendas and frameworks, and imaginations of the urban. We also see responses to the demands of programme accreditation, existing and available faculty expertise, local planning cultures, and job markets. While not explicitly articulated, some of the programmes align with globalising landscapes of planning employment, which shape the way these programmes are presented and the professional competencies they choose to highlight, as in the case of Moratuwa's development of urban management expertise, or NUS with its focus on techno-spatial analysis skills. Some programmes are distinctly catering to national demands, as seen in UNILAG, whereas others have a more regional or global objective. We also see, in the articulated imagination of the planner, a continued bias towards physical planning in most programmes, but also—in a few programmes like HKU, UCT, and AUB—a more diverse understanding of the sociopolitical nature, and multiplicity of planner's roles in practice. Some of these observations, written by other authors in individual contexts (Mahadevia and Bhatia 2018), become very distinct when comparing the universities and geographies.

The shift is also more apparent in the vision statement/objective of the programme than in the actual design of the curriculum itself. There appears to be a loss in translation between intent and proposed content; while there is recognition of contemporary and contextualised issues and needs, this is articulated better in the vision and objectives than in the modules design and selection. It may be argued that certain concerns are dealt with during course delivery, but by not spelling out key issues in the curriculum description itself, there runs a danger of critical aspects being left to the discretion of individual faculty members. This concern is again seen in how programmes engage (or not) with the issues of values, ethics, and moral choices as a planner. Some of these value questions and negotiations, as written by Taşan-Kok and Oranje (2018) and others, may be dealt with through dedicated spaces in the curriculum for reflection and debate, as is expected from the values modules of HKU or the planning seminars of AUB. In other cases, there is a possibility for a more soft-touch and integrated approach through discussion in studios or classroom-based modules. Again, the absence of dedicated spaces for such deliberations carries the risk of oversight of an issue that has been highlighted as crucial by practitioners, educators, and graduates alike. In a handful of cases where values and ethics have been given their own space in the programme, we tend to see more instrumental ways of looking at them through modules in professional practice. While it may appear reductive to look for modules focused on values and ethics—and it can be argued that they should be embedded in everything—the methods and approaches to do so within such programmes is one important line of enquiry for improving the potential of these programmes in preparing their future planners.

Kwok (1983: 93) has discussed how planning education approaches may be seen as "multi-disciplinary (which coordinates different disciplines) or inter-disciplinary (which integrates from the outset)". In this context, we see that some programmes have a great number of discrete, dedicated modules (e.g. SPA-D, UNILAG) while others are at the other end of the spectrum with far fewer, more integrated modules overall (AUB). The danger of the former type of programme is the possibility of excessive overlap or alternatively a lack of shared direction or cohesiveness across the modules, where there is less time given to negotiating or prioritising the many forms of knowing/doing and various parcels of knowledge delivered separately. Often, systemised integration across modules gives holistic "meaning" to what is learned in the programme (Kwok 1983). In this review, some programmes are clearer in their overall objectives than others; in many, they appear to be a collection of courses and the integration is not explicit. In the studio as well, which bears the potential of bringing together multiple forms of knowing, there appears to be very little change in most of the programmes, and it continues to remain an artefact of the much-criticised expert-driven or rationalist planning model. Though a few places are making an effort to run collaborative studios with communities on the ground, a number of other key stakeholders (like government and private sector actors) are still missing from the picture.

We see sustainability as a recurring term in both intent and content throughout most programmes. We also see an emphasis on physical aspects of planning such as physical infrastructure, especially in the content of modules. However, these continue to remain somewhat limited in their framing—for example, transport rather than the idea of mobility (NUS is a notable exception), or sewerage and waste management rather than service delivery, and the issues of equity and public health. Perhaps responding to prior cycles of critique on the lack of attention to implementation (Kwok 1983), and growing focus on sustainability globally, we see these included in modules dealing with development finance, and environmental management. However, apart from a few exceptions such as UCT and UNZA, other crucial themes of informality and access to land, as highlighted by UN-Habitat (2009), Watson (2011), Mahadevia and Bhatia (2018) and Denoon-Stevens et al. (2020), still appear to remain peripheral in many programmes, especially those in Asia accompanied by a long-standing lack, or very narrow training in participatory engagement. Necessary soft skills of communication and language, facilitation, and negotiation, also remain conspicuously absent for the most part, despite programmes speaking about "multi-stakeholder" engagement and planners working in "multidisciplinary teams". Instead, across all programmes, we see much weight given to computer and design software literacy, methods of analysis and research, and technical report writing—what could be referred to as the hard skills of planning. The question this raises for us is why are some concepts more explicitly included than others: for example, environment and sustainability, but land markets, informality, or equity and inclusion less often? Why do "hard" skills overshadow "soft" skills to such an extent? It is worth investigating further exactly how and why such choices are made in curricula.

This review of the ten planning programmes presents us with evidence of rich parallels and variations in the curricula within and across the two continents. It is clear that multiple influences are at play in shaping the planning education in these contexts, which merit further investigation and discussion. The review also highlights the varied attempts that are being made in different places to respond to the contemporary urban challenges within their local and regional contexts. This presents an immense opportunity for learning and innovation through South-South exchanges and partnerships—a critical, yet under-explored area for collaboration when compared with existing North–South knowledge exchange partnerships.

Acknowledgements We are very grateful to the anonymous reviewers for valuable comments and suggestions. We wish to thank Rekha Raghunathan at the Indian Institute for Human Settlements for careful reading and editing of the first draft of the paper. A special thanks to Dr. Genet Alem Gebregiorgis at TU Dortmund for constant support throughout the process of paper writing. This article is part of the research programme "Knowledge in Action for Urban Equality" (KNOW), supported by UKRI through the Global Challenges Research Fund GROW Call. Grant Ref: ES/P011225/1.

References

Adamson B, Morris P (2014) Comparing curricula. In Bray M, Adamson B, Mason M (ed) Comparative education research. Springer International Publishing, pp 309–332. https://doi.org/10.1007/978-3-319-05594-7_11

Ali AK, Doan PL (2006) A survey of undergraduate course syllabi and a hybrid course on global urban topics. J Plan Educ Res 26(2):222–236. https://doi.org/10.1177/0739456X06291500

AUB (American University of Beirut) (2018) Graduate Catalog 2018–19, Department of Architecture, pp 328–329. Available from https://www.aub.edu.lb/Registrar/Documents/catalogue/graduate18-19/architecture-mupp-mud.pdf. Accessed 8 Oct 2020

Blair N, Manda M (2016) A partnership-based model for embedding employability in urban planning education. In: Urban, planning and transport research 4. Routledge, pp 46–63. https://doi.org/10.1080/21650020.2016.1145070

Bray M, Adamson B, Mason M (2014) Comparative education research: approaches and methods. Comparative Education Research Centre, University of Hong Kong, Hong Kong

Butt A, Ratnayake R, Budge T (2013) Planning education and inter-cultural collaboration: awareness, innovation, reflection and preparation for practice. Bhumi, Plan Res J 3:1–10. https://doi.org/10.4038/bhumi.v3i1.10
Chatterji T, Soni A (2016) Chapter 5: positioning urban governance in planning pedagogy. In: Kumar A, Meshram DS, Gowda K (ed) Urban and regional planning education: Learning for India.Springer Science+Business Media, Singapore, pp 61–77
Davoudi S (2015) Planning as practice of knowing. Plan Theory 14:316–331. https://doi.org/10.1177/1473095215575919
Denoon-Stevens SP, Andres L, Jones P, Melgaço L, Massey R, Nel V (2020) Theory versus practice in planning education: the view from South Africa. In: Planning practice and research. Routledge, pp 1–17. https://doi.org/10.1080/02697459.2020.1735158
Diaw, K, Nnkya T, Watson V (2002) Planning education in sub-Saharan Africa: responding to the demands of a changing context. In: Planning practice & research 17. Routledge, pp 337–348. https://doi.org/10.1080/026974502200005689
Edwards MM, Bates LK (2011) Planning's core curriculum: knowledge, practice, and implementation. J Plan Educ Res 31(2):172–183. https://doi.org/10.1177/0739456X11398043
Flyvbjerg B (2006) Five misunderstandings about case-study research. Qual Inq 12(2):219–245. https://doi.org/10.1177/1077800405284363
Frank AI (2018) Chapter 9: adapting, shifting, defining new roles: education for a maturing professional field. In: Frank A, Silver C (ed) Urban planning education: beginnings, global movement and future prospects. Springer International Publishing AG, pp 131–145
Frank A, Silver C (2018) Urban planning education: beginnings. The Urban Book Series. Springer International Publishing AG, Global movement and future prospects. https://doi.org/10.1080/09640566208730631
Friedmann, J (1996) The core curriculum in planning revisited. J Plann Educ and Res 15(2):89–104.
Hague, C (2001) The formation of the Commonwealth Association of Planners Network of Planning Schools: a response to the Istanbul+5 development agenda. In: World planning schools Congress, Shanghai
HKU (University of Hong Kong) (2020) Master of science in urban planning. https://www.arch.hku.hk/programmes/upad/master-of-science-in-urban-planning/. Accessed 8 Oct 2020
Hou L (2018) Chapter 6: six decades of planning education in china: those planned and unplanned. In: Frank A, Silver C (ed) Urban planning education: beginnings, global movement and future prospects. Springer International Publishing AG, pp 81–99
Kumar A, Meshram DS, Gowda K (2016) Urban and regional planning education: learning for India. Urban and Regional Planning Education.Springer Science+Business Media, Singapore
Kunzmann KR (2015) The state of the art of planning and planning education in Asia. Disp—Plan Rev 51:42–51. https://doi.org/10.1080/02513625.2015.1134961
Kunzmann KR, Koll-Schretzenmayr M (2015) Unchartered territory: planning and planning education in Asia 2015. Disp–plan Rev 51:20–21. https://doi.org/10.1080/02513625.2015.1134925
Kusiima A (2008) Makerere University, Uganda: Urban Planning Programme. AAPS Conference 1–24.
Kwok R-W (1983) Education in urban planning. Habitat Int 7:91–101. https://doi.org/10.1016/0197-3975(83)90014-0
Mahadevia D, Bhatia N (2018) Urban planning education for inclusive cities: global and Indian perspectives
McFarlane C (2010) The comparative city: knowledge, learning, urbanism. Int J Urban Reg Res 34(4):725–742. https://doi.org/10.1111/j.1468-2427.2010.00917.x
Meshram DS, Meshram S (2016) Chapter 2: energizing planning education in India. In: Kumar A, Meshram DS, Gowda K (ed) Urban and regional planning education: learning for India. Springer Science+Business Media, Singapore, pp 17–32
NIUA (National Institute of Urban Affairs). 2017. Recommendations for an improved master's level urban planning curriculum. Delhi, India

Nnkya T, Lupala J (2008) Ardhi University, Dar Es Salaam, Tanzania: School of Urban and Regional Planning. AAPS Conference 1–21

Nordtveit B (2016) Trends in Comparative and international education: perspectives from the comparative education review. Ann Rev Comp Int Educ 27–37. https://doi.org/10.1108/S1479-367920160000030001

NUS (National University of Singapore), School of Design and Environment (2020) 4.2.2.9 Master of urban planning. http://www.nus.edu.sg/nusbulletin/school-of-design-and-environment/graduate-education/coursework-programmes/degree-requirements/master-of-urban-planning/. Accessed 8 Oct 2020

NUS (National University of Singapore), Department of Architecture (2020) Master of urban planning. http://www.sde.nus.edu.sg/arch/programmes/master-of-urban-planning/. Accessed 8 Oct 2020

Oborn P, Walters J (2020) Planning for climate change and rapid urbanisation: survey of the built environment professions in the Commonwealth

Odendaal N (2012) Reality check: planning education in the African urban century. Cities 29. Elsevier Ltd, pp 174–182. https://doi.org/10.1016/j.cities.2011.10.001

Oduwaye L, Lawanson TO (2008) University of Lagos, Nigeria: Department of Urban and Regional Planning. AAPS Conference 1–33

Oduwaye L, Olajide O (2012) Incorporating informality into urban and regional planning education curriculum in Nigeria. Town Reg Plan 60:31–37

Peña S (2019) Urban and regional planning education in Mexico. J Plan Educ Res. https://doi.org/10.1177/0739456X19889662

Prakash P (2016) Ethics and planning education in India. In: Kumar A, Meshram DS, Gowda K (ed) Urban and regional planning education: learning for India, pp 181–192. https://doi.org/10.1007/978-981-10-0608-1_13

Ratnayake R, Butt A (2017) Encounters with the unfamiliar: international planning education. Int Plan Stud 23:51–64. https://doi.org/10.1080/13563475.2017.1339347

Robinson J (2011) Cities in a world of cities: the comparative gesture. Int J Urban Reg Res 35(1):1–23. https://doi.org/10.1111/j.1468-2427.2010.00982.x

Robinson J (2015) Thinking cities through elsewhere. Prog Hum Geogr 40(1):3–29. https://doi.org/10.1177/0309132515598025

Sen S, Umemoto K, Koh A, Zambonelli V (2016) Diversity and social justice in planning education: a synthesis of topics, pedagogical approaches, and educational goals in planning syllabi. J Plan Educ Res 37(3):347–358. https://doi.org/10.1177/0739456X16657393

Short EC (1991) Forms of curriculum inquiry. State University of New York Press, Albany, NY

Siame G (2016) The value and dynamics of community-based studio projects in planning education in the Global South. Berkeley Plan J 28:40–67. https://doi.org/10.5070/BP328133858

SPA-D (School of Planning and Architecture, New Delhi) (2020) About the Department of Urban Planning. Available from http://spa.ac.in/User_Panel/UserView.aspx?TypeID=1429. Accessed 8 Oct 2020

Stiftel B, Mukhopadhyay C (2007) Thoughts on Anglo-American hegemony in planning scholarship: Do we read each other's work? Town Plan Rev 78:545–572. https://doi.org/10.3828/tpr.78.5.2

Taşan-Kok T, Oranje, M (2018) Why it is important to give voice to young practitioners. In: Taşan-Kok T, Oranje M (ed) From student to urban planner: young practitioners' reflections on contemporary ethical challenges. Routledge, New York, pp 1–11. https://doi.org/10.4324/9781315726854-1

Todes A, Harrison P, Watson V (2003) The changing nature of the job market for planning in South Africa: implications for planning educators. Town Reg Plan 46:21–32

UCT (University of Cape Town) (2020) Bachelor of City Planning Honours degree and Masters of City and Regional Planning degree (linked). http://www.apg.uct.ac.za/apg/city-and-regional-planning. Accessed 8 Oct 2020

United Nations (2017) New urban agenda. https://habitat3.org/wp-content/uploads/NUA-English.pdf. Accessed 8 Oct 2020
UNDESA (United Nations Department of Economic and Social Affairs) (2018) World urbanization prospects 2018. United Nations, New York
UN-Habitat (2009) Global report on human settlements 2009: Planning sustainable cities.
UNILAG (University of Lagos) (2017) Departmental Handbook 2017–2020: Department of Urban and Regional Planning. Faculty of Environmental Sciences. University of Lagos, Lagos
University of Moratuwa (2020) Student's handbook—Master of Spatial Planning, Management & Design—2016/18: Department of Town & Country Planning: 19
UNZA (University of Zambia) (2018) Postgraduates: overview. https://www.unza.zm/academics/postgraduate-programmes/master-of-science-spatial-planning. Accessed 8 Oct 2020
van Horen B, Leaf M, Pinnawala S (2004) Localizing a global discipline: designing new planning programs in Sri Lanka. J Plan Educ Res 23(3):255–268. https://doi.org/10.1177/0739456X03261282
Watson V (2008) University of Cape Town, South Africa: City and Regional Planning Programme. AAPS Conference 1–19
Watson V (2009) Seeing from the South: refocusing urban planning on the globe's central urban issues. Urban Stud 46:2259–2275. https://doi.org/10.1177/0042098009342598
Watson V (2011) Inclusive urban planning for the working poor: planning education trends and potential shift. WIEGO Working Paper (Urban Policies). Vol. 21. WIEGO Working Papers
Wesely J, Allen, A (2019) De-colonising planning education? Exploring the geographies of urban planning education networks. Urban Plan 4:139–151. https://doi.org/10.17645/up.v4i4.2200

Geetika Anand is a doctoral researcher at the University of Cape Town (UCT), focusing on practice biographies of urban development practitioners in South Africa. Working with the Indian Institute for Human Settlements (IIHS) since 2010, she is currently a researcher on the Knowledge in Action for Urban Equality (KNOW) programme, looking at planning education in Asia and Africa. Trained in urban planning, Geetika has 15 years of experience in the sector spanning the areas of master plans, housing, water and sanitation. She has a special interest in field-based learning and is part of studio teaching teams at SPA, the School of Planning and Architecture in New Delhi, IIHS, and the UCT

Nandini Dutta is an architect and planner from Auroville, practising since 2016. She has developed children's 'sensory' gardens for CCAMH-Caritas in Cambodia, and for Chengalpattu Medical College Government Hospital in India. Since May 2019, she has been a research consultant with the Indian Institute of Human Settlements in Bangalore. Her work is part of the Global Challenges Research Fund (GCRF) project, Knowledge in Action for Urban Equality, and focuses on planning education and curricula in the Asian and African regions. She also contributes to research for sustainable planning with the Max Lock Centre in New Delhi under the development research and consultancy unit of the University of Westminster.

Chapter 6
Climate Change Adaptation and Planning Education in Southern Africa

Abraham R. Matamanda, Jennilee M. Kohima, Verna Nel, and Innocent Chirisa

Abstract Cities in Southern Africa are experiencing a rapid rate of urbanisation, which exacerbates the impacts of climate change on cities. The recent droughts and water stress in Cape Town, South Africa and Windhoek, Namibia, impacts of Cyclone Idai that destroyed 90% of Beira city, and recurrent heatwaves are evidence of the impacts of climate change on cities in the region. Planners are responsible for the spatial configuration of spaces and places such that cities are safe, resilient, sustainable, and inclusive; hence planning for climate change is imperative. In this study, we argue that the recurrence of climate change-related disasters in Southern Africa reflects the lack of skills, knowledge and capabilities among planners to integrate climate change adaptation into urban planning processes. Like any other profession, planning practice is informed by education and training of the graduates, which influences their worldview and ideology that they take into the professional world. This study examines the contribution of planning education to climate change adaptation in Southern Africa, using the case of Zimbabwe, South Africa and Namibia. Using content analysis of course syllabi in terms of the pedagogy on climate change adaptation, the study identifies the knowledge, skills, and abilities schools impart to planning students. The study reveals that climate change is recognised as a planning dilemma, but it is yet to be integrated into the planning curriculum and is consequently marginalised.

A. R. Matamanda (✉)
Department of Geography, University of the Free State, Bloemfontein, South Africa
e-mail: matamandaa@gmail.com

J. M. Kohima
Department of Architecture and Spatial Planning, Namibia
University of Science and Technology, Windhoek, Namibia

V. Nel · I. Chirisa
Department of Urban & Regional Planning,
University of the Free State, Bloemfontein, South Africa
e-mail: NelVJ@ufs.ac.za

I. Chirisa
Department of Demography Settlement & Development,
University of Zimbabwe, Harare, Zimbabwe

G. Alem Gebregiorgis et al. (eds.), *Planning Cities in Africa*, The Urban Book Series,
https://doi.org/10.1007/978-3-031-06550-7_6

Keywords Urbanisation · Climate change · Adaptation · Planning education · Case study

6.1 Introduction

Climate change impacts on cities have long been recognised as cities increasingly become hotspots for different climate-induced hazards and disasters (Baarsch and Schaeffer 2019). In recent years, the persistent droughts that compromised water security in Cape Town, South Africa, indicate the impacts of climate change on African cities (Scott et al. 2019). In 2019, Cyclone Idai was associated with heavy rains, flooding and landslides that caused immense damage to property, livelihoods and human life across cities in Malawi, Mozambique and Zimbabwe. In Mozambique, 90% of the city of Beira was destroyed, which amounted to millions of dollars lost in infrastructure and services (Shaban 2019). In other cities, heatwaves have become the norm and have been described by Carter (2018) as "a potential silent killer of citizens in African cities". New diseases and plagues that destroy farmlands have emerged that jeopardise food security, as recently experienced in East Africa (Stone 2020). The events from climate change-induced floods in Durban and destroyed some planned suburbs raise critical questions for professional planners as to how they can best adapt settlements to climate change. Against this background of increasing climate-related hazards and disasters, African cities are rapidly urbanising (Obasanjo et al. 2020). Therefore, planners must adapt cities to climate change to increase the resilience, sustainability and safety of these spaces where an increasing share of the continent's population will be residing soon.

The role of urban planning in adapting cities to climate has long been acknowledged where climate change is identified as a "super wicked problem" that requires extraordinary skills from planners (Campbell 2006; Lyth and de Chastel 2007; Broto 2014). What remains under-researched from the African context is how planners are trained and equipped to address emerging problems induced by climate change in their professional practice (Davidson and Lyth 2012). Albeit the implications of climate change prevalent in cities, Adelina, et al. (2020) note that it appears that planning education in Africa has not been scaled-up to systematically integrate climate change issues. Watson and Odendaal (2012) have recognised that African cities have changed dramatically and have different problems to combat, including climate change. Hence the need for a change in planning education to address these contemporary challenges.

The need for this change was recognised by participants at the 2010 Association of African Planning Schools (AAPS) Conference where the integration of climate change in planning schools was identified among the key themes, and planning schools were called upon to critically think about enhancing the planning curriculum (AAPS 2010). The major drawback in the existing planning curriculum, which, according to Odendaal (2012), fails to integrate many contemporary planning issues in African cities. The curricula and educational philosophy remain entrenched in the colonial system. Odendaal (2012: 175) has lamented that "the planning education philosophy and pedagogical approaches in most African planning schools are inherited from the colonialists and in many instances reflect a mismatch between plans, urban space and training of planners". Planners, who are at the centre of land development processes, are responsible for the spatial ordering of cities and, ultimately, the functioning and form of the city (Giordano et al. 2020).

We explore how planning education in selected planning schools integrate pressing climate change issues based on this background. We undertake a situation analysis and needs assessment of the existing skills, knowledge and capabilities learned in planning education about climate change adaptation, resilience and mainstreaming. We ask the question: How are planners trained to respond to climate change adaptation? Specifically, the study includes the following objectives:

- Identify the modules that focus on climate change.
- Examine the extent to which the identified modules reflect on climate change adaptation.
- Explore the knowledge, skills and abilities relating to climate change adaptation in cities that planning schools are imparting to planning students.
- Assess the scope and content of the curriculum and guidelines for selected modules with regards to climate change adaptation in African cities.
- Analyse the gap in planning education concerning climate change adaptation for African cities.
- Outline lessons for planning education concerning the integration of climate change in African cities.

6.2 Climate Change and Planning Education: A Literature Review

African cities are complex spaces confronted with multiple socio-economic and political challenges (Zellner and Campbell 2015). Climate change adds to these challenges. The realities of climate change for cities and towns have increased the urgency for planners and other professionals to adapt cities to these realities (Blakely 2007). The impact on human settlements, especially the most vulnerable spaces that include informal settlements that the poor inhabit, call for immediate action to promote the liveability of these settlements (Satterthwaite et al. 2020).

It is also interesting that it is not only informal settlements where the poor live that are vulnerable to climate change impacts. The increasing water scarcity in African cities, although arising from other factors such as poor governance, cannot be detached from the effects of climate change. An example is Cape Town in 2018, when the city went dry (Scott et al. 2019). Destruction of key infrastructure as happened in Beira, the burden on taxpayers and the need for adaptation strategies require planning professionals who understand the complex dynamics of climate change and the consequences of the process (Davidson and Lyth 2012). Therefore, it becomes critical to understand how planners are educated about climate change.

The place of planning education in climate change mitigation and adaptation cannot be underestimated. Scholars are increasingly articulating the nexus between climate change, and university education (Mochizuki and Bryan 2015; Leal Filho 2017; Monroe et al. 2019), resulting in a growing body of knowledge. This has been emphasised internationally through Article 12 of the Paris Agreement that among other issues, encourages governments to enhance climate change education and training (UNFCCC 2015). Moreover, universities are encouraged to create new academic programmes in various disciplines to ensure that future professionals understand the challenges posed by climate change and have the necessary tools to effect successful mitigation and adaptation (UNESCO 2017).

Planning education may be positioned on the first two of the four pillars of education, as shown by Delors (1996). Firstly, Delors (1996) indicated that learning might be undertaken with the objective of knowing—learning to know. In this instance, students are provided with theoretical content that equips them with the knowledge to understand climate change issues and dynamics and how they manifest in cities. The rationale is to ensure that students appreciate the causal factors and the consequences of climate change from a planning perspective. Molthan-Hill et al. (2019) explain that through learning to know, students are also introduced to the tools and strategies that may be used to address different climate change impacts on cities.

Secondly, according to Delors (1996), students may be taught to learn to do so that they can apply their knowledge in a real-world context. This type of learning is the ideal form of education through which skills are turned into competencies. Rather than solely acquiring knowledge, students are trained and educated to become competent professionals with coping skills and the ability to adapt to different situations. This is critical considering that climate change and cities are complex. Hence planners must be trained to understand systems and envision different solutions and future scenarios using essential skills (Molthan-Hill et al. 2019). In this regard, planning education becomes critical and may be enhanced through environmental design, land use planning and environmental planning modules. Studio type of education may also be relevant as it helps bring students in sync with local realities.

Climate change education in planning schools is often based on four approaches outlined by Molthan-Hill et al. (2019). Firstly, climate change education may be through piggybacking, a common approach used in most institutions where some components of climate change are integrated into existing modules or courses. Molthan-Hill et al. (2019) explain that the existing structure of the module or course is not changed; rather, some climate change-related content is added. Secondly, climate change learning may be through mainstreaming, which still involves the integration of climate change content in existing modules and courses, but this is done with a broader cross-curricular perspective. For example, the University of Dar es Salam in Tanzania mandates all students to obtain at least a basic understanding of climate change. Thirdly, there may be specialisation by designing specific climate change-related modules and courses. Climate change may thus be made an elective course that some students will focus on, while others focus on project management, transport planning or infrastructure planning. Lastly, climate change learning could be implemented through an inter-disciplinary approach adopted such that the modules on climate change may be offered across a particular university or faculty. Thus, climate change is addressed from a multi-disciplinary perspective to allow students to appreciate its complexity and ultimately be better equipped to address its impacts (Molthan-Hill et al. 2019). This may be done so that students appreciate the socio-economic, scientific and political dimensions of climate change.

6.3 Methodology

The study was qualitative and adopted a case study design. Secondary data collection methods were used in this study and included a review of planning courses from selected universities in Southern Africa. Three planning schools were purposively selected to inform this study, namely the University of the Free State (UFS), South Africa; University of Zimbabwe (UZ) and Namibia University of Science Technology (NUST), Namibia. The University of Zimbabwe was selected, because it recently reviewed its curriculum. Hence, it was critical to identify the outcomes of such an assessment and what recommendations were made regarding including climate change content in the curriculum. The modules that focused on climate change, their content and the scope and breadth of the covered issues were determined. Gaps in the planning courses were identified. Using content analysis of course syllabi in terms of the pedagogy on climate change adaptation, the study identified the knowledge, skills and abilities that planning schools imparted to planning students.

6.4 Case Studies

6.4.1 South Africa: University of the Free State

South African cities are vulnerable to climate change, the effects of which are flooding, heatwaves and sea-level rise. The impacts of climate change have resulted in water scarcity in cities such as Cape Town. In contrast, the constantly expanding human settlements in Alexandra, Johannesburg, Cape Town and Durban have been exposed to climate change-induced flooding. The destruction of ecosystem services cannot be understated considering the green initiatives that are imperative to redress climate change.

The Department of Urban and Regional Planning (DURP) at the University of the Free State is one of the planning schools in South Africa that train planners who are eligible to register as professionals with the South African Council for Planners (SACPLAN)[1] once they have graduated and undertaken the requisite candidacy period (internship). The DURP strives through excellent teaching and scholarship to deliver competent urban and regional planners and research that will contribute to creating sustainable human settlements and improved quality of life, particularly in Africa. The philosophy of the DURP is premised on the need to produce proficient planners who can make a difference to their environment. In this regard, the department has resolved to place "sustainable human settlements" among its research focus. The thrust of this has been on the green agenda with a focus on sustainability, climate change adaptation and mitigation, and resilience as the substantive issues that students should be acquainted with.

To register as a professional planner with SACPLAN, graduates from DURP must pass the Bachelor of Spatial Planning Honours (BSPH) and the Master's in Urban and Regional Planning (MURP). Once the graduates have been awarded their qualifications, they must complete the necessary practical training stipulated by SACPLAN. The BSPH forms the foundation for professional qualification in the DURP and involves nine modules that cover different aspects of spatial planning. Although there is no single module in the DURP programme that specifically focuses on climate change, climate change is covered in the following modules:

- Research in Environmental Planning,
- Human Settlement Planning,
- Research in Regional Planning Theory,
- Research in Urban Development,
- Basic Practice.

At the Honours level, students are expected to understand the dynamics of climate change and how it is a planning issue. In most modules, several cross-cutting issues such as climate change and informality pertinent to developing sustainable settlements are integrated into the curriculum. The Research in Environmental Planning module addresses ecological and environmental issues in settlement planning and development and includes climate change issues. The other modules mentioned include measures and proposals to deal with climate change. Hence, climate change issues have been integrated into the existing curriculum and become an issue that students are cognisant about in what Delors (1996) calls learning to know. The BSPH also trains the students to apply their knowledge in the real world (Delors 1996). This is confirmed by the objective of the BSPH that seeks to capacitate graduates so that they can practically apply theory in urban and regional planning projects. This is demonstrated by an assignment in the regional planning module that requires students to identify critical environmental issues such as climate change and suggest ways of preventing, mitigating and managing the problems.

[1] The SACPLAN is the statutory council established within the Planning Profession Act, 2002 (Act 36 of 2002) whose sole purpose is to regulate the planning profession based on certain terms and conditions.

The MURP also has various modules that adopt the piggybacking approach to climate change education. Climate change is integrated into modules that include geographic information systems for planners. The focus of the modules is to capacitate learners with technical skills critical in climate change intervention, as mapping usually enables the understanding of the impacts of certain hazards such as floods. In this regard, one of the assignments for this module, tasked students to "Create a map that contains a Spatial Development Framework proposal for Mangaung, as one of Mangaung's proposals, entails reinforcing climate policy objectives through green interventions and protecting the environment". In the Theory of Planning module, students are introduced to different planning theories and emerging concepts in urban planning. Climate change features as a cross-cutting issue, which is discussed in one of the study units where students are familiarised with different substantive issues in planning, such as resilience and sustainability. This particular module focuses on raising awareness of the impacts of climate change. Moreover, the Urban Research Project module also addresses issues in climate change as it identifies it as a critical urban issue that requires the attention of planners. For example, an assignment the students were given read: "effectively link economic, environmental and social aspects in relation to sustainable urban planning and urban management: explain how these issues, as well as climate change, have been integrated into the SDF (Spatial Development Framework) of Mangaung".

6.4.2 Zimbabwe: University of Zimbabwe

Urban and Regional Planning, as a discipline, is currently offered at the UZ, Great Zimbabwe University (GZU), and Lupane State University (LSU). In 1984, the Department of Rural and Urban and Regional Planning (DRUP) was formed, supported by the government, at the University of Zimbabwe and the Institute of Social Studies (ISS) in the Netherlands, which was willing to fund research and the teaching in aspects of rural development within the Faculty of Social Studies. The initial core areas of research and teaching in the DRUP were in 1988, when the department had its first undergraduate enrolments. At the time, the degree programme was centred on modules consisting of a combination of aspects of sociology, urban design, architecture, planning, real estate, quantity surveying, engineering, project planning and management, environmental systems, transport planning and management, and professional planning.

The foregoing historical background to the planning discipline and profession in Zimbabwe is necessary to understand how the subject of climate change has either been absorbed or ignored in certain courses/modules of the curriculum. The courses can be clustered in five categories:

- *Environment-oriented cluster* begins with understanding environmental processes and systems and tapers towards why planners should understand how their

designs impact the environment and how the environment (climate, geology, geomorphology, soils, ecology and hydrology) may impact on the designs.
- *Design-oriented cluster*, which covers aspects of urban and environmental design theory and entails studio design: the design of the outdoor to include both soft and hard landscapes, adequate stormwater drainage, the inclusion of trees not only for ornamental purposes but functional purposes such as shade and reduction of floods.
- *Circulation and service-oriented cluster*, focusing on how services (water, sewer, transport, power and construction) within a planned system environment are provided, managed and disposed of.
- *Theoretically oriented cluster*, which seeks to offer multi-perspectivity as to how and why certain development happens or fails to happen in spaces with or without plans. The various perspectives are sub-clustered in environmental, engineering, philosophical, political, economic, technological and professional, to name but a few. In this cluster, planning theory, planning thought and professional practice feature directly as modules that seek to offer explanations and achieve understanding.
- *Research and policy cluster* acknowledges that planning as a continuous process thrives with emerging data generated by responding to a policy environment created by local, central and international governing bodies.

This clustering is not, per se, a smooth and simply defined parameterisation as there are overlaps between them. In creating a graduate planner, several of the courses emphasise the subject of climate change differently while others become almost silent on the subject. When layout planning is done in the design-oriented cluster, the emphasis on climate is quite clear and obvious. For example, in designing road width, the road should be wide enough to accommodate stormwater drains, which can be used to channel water as quickly as possible. Road designs should also allow for soft landscaping that acts as a sustainable urban drainage system, enhancing or allowing for natural drainage to replenish underground water sources.

Climate change features as a subtopic in the Environmental Systems, Environmental Planning, Urban Development and Management and Urban Policy and Planning courses offered by the Zimbabwean universities. Climate change is perceived as already affecting Zimbabwe and its urban areas. Urban and regional planning greatly influences some of the critical climate change mitigation strategies, such as energy efficiency at different scales, sustainable urban mobility (walkability and mass transportation) and carbon sinks through green infrastructure. Urban and regional planning also plays a vital role in building resilience in climate change by creating a resilient urban environment, which can be through promoting density, diversity and connectivity in urban areas.

In Project Planning and Management, Quantitative Techniques and Planning Techniques and Methods, climate change is barely seen in the curricula. Lecturers acknowledged that there is little awareness of climate change and its impacts on the general population. They acknowledged that local authorities should be at the forefront in promoting climate adaptation and mitigatory measures in cities. However, little is known about the efforts that these municipalities are taking.

In the teaching of Quantitative Techniques, it is possible to employ quantitative techniques in decision-making and use relevant illustrations in linear programming. The same can also be said for the teaching of Planning Techniques and Methods, as research methods used in planning are part of the aspects covered. The teaching of Project Planning and Management involves students understanding the project planning cycle and project management techniques. Projects are affected by climate change; hence there is a need to understand the application of climate adaptive projects. Therefore, there is an apparent need to include climate change aspects in the course.

In the teaching of the ArcGIS and Remote Sensing mapping in planning and development, climate change is acknowledged as a phenomenon that needs to be well understood as it affects all facets of life and needs to be factored into all planning and development phases. It is an issue, because all plans for development and management should encompass ways of adapting to climate change, reducing the negative impact of climate change, heightening positive impact and developing without accelerating climate change. Understanding the occurrence and effects of climate change through change analysis from high-resolution remote sensing images using different ArcGIS software is essential.

The development planning module stresses the master plans, development plans, sectoral development and development change. There is a strong relationship between climate change and development. Climate change is a game-changer in the zonation of land uses and design of buildings, especially in urban areas.

Planning techniques and methods, project planning and development, development planning, research methods and statistics, development and governance, property management and property valuation should all consider the influence of climate change. Climate change is both a planning and real estate issue. It is a planning issue because planning does not occur in a space or a vacuum, but in an environment influenced by climate change. For example, the physical environment has changed because of climate change, which has implications for spatial planning (place-making). It is a real estate issue in the sense that the way buildings are designed needs to consider issues of sustainability, which includes energy saving in the form of green buildings. This also implies that the materials to be used should be environmentally friendly and meet climate change demands. Nevertheless, some students view climate change as an area to be studied mainly in the department of geography. Some may not understand the implications of climate change to the planning fraternity.

6.4.3 *Namibia: Namibia University of Science and Technology*

Namibia is considered one of the countries most vulnerable to climate change, due to its very dry semi-arid to desert nature (Reid et al. 2008: 453; Tervo-Kankare et al. 2018: 274). The frequency of droughts will thus increase with predictions of water shortages in urban areas. The lack of or limited financial resources, skills and technologies, and high levels of poverty contribute to the vulnerability of Namibia to the realities of climate change. Thus, Namibia developed a national policy framework to mitigate and adapt to the adverse effects of climate change on the environment. Like many cities of the developing world, the rapidly growing city of Windhoek battles with the impacts of climate change such as rising temperatures, falling dam levels and erratic rainfall, which has led to both droughts and flash floods (Van Rensburg 2016: 623; Scott et al. 2018: 6). The high level of urbanisation and the proliferation of informal settlements in the city are possible contributing factors. Windhoek is expected to face even greater droughts, floods, biodiversity loss and water supply depletion. Hence, the city of Windhoek developed its Integrated Climate Change Strategy and Action Plan to facilitate its response to climate change (Stockholm Environment Institute 2018: 36–37). This strategy is developed in line with and to support the National Policy on Climate Change and the National Climate Change Strategy and Action Plan.

The Department of Architecture and Spatial Planning at the Namibia University of Science and Technology (NUST) was established upon the transformation of the university from the Polytechnic of Namibia. The Department of Architecture was formerly under the School of Engineering. The Section of Land Use Planning was formerly under the Department of Land Management, which was created in 2012. It is the only department offering planning education in Namibia and focuses on an integrative, transdisciplinary understanding of the urban environment. It is considered as the driver of change to overcome apartheid planning. It also revised local standards and regulations to align with contemporary issues and challenges by providing the industry with relevantly skilled, knowledgeable and competent young professionals. The department has seen a steady increase in the student population interested in the profession for rural and urban development and planning streams. The Town and Regional Planning Honours degree offered at NUST is professionally accredited by Namibia Council of Town and Regional Planners and Namibia Institute of Town and Regional Planners. These two bodies extensively influence the content of courses offered in spatial planning.

The department understands its important position as the only planning school in Namibia and the requirement to create and instill awareness and understanding of climate change in the built environment among the students. It offers various graduate courses focusing on environmental and climate-related research and teaching in its programme. Climate change is highly integrated into the curricula. The Advanced Theory of Urban and Regional Planning course includes a unit on climate change. The Environmental Planning course covers multiple topics such as air pollution problems and solutions, climate change causes, evidence, trends and solutions at various levels. Sustainable Settlement Planning entails exploring the blue and green economies for sustainable planning. To graduate multi-skilled built environment professionals with the town and regional planning and regional and rural development specialisations, the curriculum for Town and Regional Planning focuses on teaching climate change implications in land use, urban resilience, sustainability, environment and development. The Regional and Rural Development curriculum looks more into understanding and solving the complex problems resulting from climate change through learning practically and theoretically about climate change.

Over the years, planning educators at NUST observed the depth of students' knowledge about climate change adaptation in cities through class dialogues, discussions and participation. The sophistication of analysis while exploring climate change adaptation in cities is evident through assessments. Educators referring to SDG Goal 11 "*Make cities inclusive, safe, resilient and sustainable*" and other development and climate change goals ensure that students apply critical thinking and problem solving while reacting to climate change challenges of urban areas in Namibia and Africa.

The current curriculum offers at least 22 out of 82 courses on climate change and climate change adaptation to a greater and lesser extent through its undergraduate and postgraduate programmes. This translates to 27% of the Bachelor and Bachelor Honours in Town and Regional Planning courses. Bachelor Honours of Regional and Rural Development includes climate change-related content as well. Courses such as Integrated Environmental Management, Environmental Planning and Natural Resources, and Sustainable Development offered to undergraduates cover climate change and related thematic areas extensively within the local and global context. Additionally, the students are encouraged to consider research topics linked to climate change, adaptation and mitigation and sustainability, through courses such as Planning Theory, Sustainable Local Economic Development, Rural Development Methods and Practice, Spatial Planning Practice and Mini-theses. Students in the department have already produced mini-theses with climate change topics in both streams over the past years.

Through the 22 courses identified as covering climate change adaptation directly or indirectly, students are introduced to intermediate knowledge, skills and abilities to contribute to a climate change adaptation discussion on urban areas. There are 14 courses, seven courses each within the programmes of Town and Regional Planning and Regional and Rural Development. Climate change is covered in three courses in both the Town and Regional Planning Honours and the Regional and Rural Development Honours. They cover introductory information on climate change in the global and Namibian context, then move on to impacts, adaptation and mitigation measures. They also cover the nexus between natural resources and climate change. A mini-thesis research topic from 2018 dealt with the "*Evaluation of Spatial Planning Adaptation Measures for Sea Level Rise in Langstrand, Walvisbay*".

6.5 Discussion and Conclusion

The universities discussed above demonstrate that climate change is recognised. It has been observed that climate change is recognised as a planning issue that students must understand its manifestation and its impacts and ways to mitigate and adapt cities and regions to the mounting problems. As part of the accreditation of any South African planning programme, SACPLAN requires among other things, that the curriculum include setting standards and competencies (SACPLAN 2014). While these competencies require planners to know and apply concepts to sustainable cities and regions, climate change is not specifically mentioned (Lewis and Nel 2020). Thus, there is no explicit requirement for South African planning programmes to stress climate change. Hence, issues related to climate change are usually subsumed into other modules, and the emphasis thereon depends on the planning school's focus. This is in line with the AAPS calls for climate change to be included in planning courses. Still, it does not feature as centrally as aspects such as informality. Instead, it is rather included in various guises within the curricula, not only in South Africa but also in the UZ and NUST.

All three planning schools recognise the realities of climate change and how it manifests as a planning issue. In this regard, climate change has been integrated into the respective curricula but mainly through a piggybacking approach. This approach involves learners being introduced to climate change and increasing their awareness of the realities of the challenges posed by this problem. There is also some effort made to equip the students with the skills to address climate change issues in real life as evident from the ArcGIS modules at the UFS and UZ, which require students to make simulations of real-life incidences and how they may inform policy. Such an approach exemplifies what Delors (1996) terms "learning to do", which betters the more passive "learning to know".

However, given the wide-ranging and increasingly devastating effects of climate change on cities, health, livelihoods and lives, climate change needs more attention within the planning curricula. The knowledge of climate change, its various impacts and relevant regional predictions must become more prominent, and more attention must be given to preventing, mitigating and coping with the effects of climate change in urban, regional and rural planning and management. This should better the existing piggybacking that characterises the existing planning education. It is suggested that climate change should be introduced as a standalone module, which is compulsory for all students. This would also require the teaching staff to be capacitated in climate change issues and necessitate them to understand the dynamism of climate change that transcends across disciplines and must be approached from different worldviews and contexts.

References

AAPS (Association of African Planning Schools) (2010) Summary of discussion and proceedings of the 2010 AAPS Conference: Revitalising planning education in Africa. 5th to 8th October 2010 Coral Beach Hotel Dar es Salaam Tanzania

Adelina C, Archer D, Johnson O, Opiyo RO. (2020) Governing sustainability in secondary cities of the Global South. SEI Report, July 2020. Stockholm Environment Institute, Stockholm

Baarsch F, Schaeffer M (2019) Climate change impacts on Africa's economic growth. Africa Development Bank, Abidjan

Blakely EJ (2007) Urban planning for climate change. Lincoln Institute of Land Policy, Cambridge, MA

Broto V (2014) Planning for climate change in the African city. Int Dev Plan Rev 36(3):257–264

Campbell H (2006) Is the issue of climate change too big for spatial planning? Plan Theory Pract 7(2):201–230

Carter S (2018) Heatwaves could become a silent killer in African cities. African Climate & Development Initiative, 30 November. Available at http://www.acdi.uct.ac.za/news/heatwaves-are-silent-killer-african-cities. Accessed 21 July 2020

Davidson J, Lyth A (2012) Education for climate change adaptation—enhancing the contemporary relevance of planning education for a range of wicked problems. J Educ Built Environ 7(2):63–83

Delors J (1996) Learning: the treasure within. Report to UNESCO on the International Commission on Education for the Twenty-first Century. UNESCO, Paris

Giordano R, Pilli-Sihvola K, Pluchinotta I, Matarresse R, Perrels A (2020) Urban adaptation to climate change: climate services for supporting collaborative planning. Clim Serv. https://doi.org/10.1016/j.cliser.2019.04.004

Leal Filho W (2017) Climate change research at universities addressing the mitigation and adaptation challenges. Springer, Berlin

Lewis M, Nel V (2020) Setting standards and competencies for planners. In: Silva CN (ed) Routledge handbook of urban planning in Africa. Routledge, London/New York, pp 162–176

Lyth A, de Chastel L (2007) Shifting towards sustainability: education for climate change adaptation in the built environment sector. Australian Planner 44(3):12–14

Mochizuki A, Bryan A (2015) Climate change education in the context of education for sustainable development: rationale and principles. J Educ Sust Develop 9:4–26

Molthan-Hill P, Worsfold N, Nagy GJ, Leal Filho W, Misfud M (2019) Climate change education for universities: a conceptual framework from an international study. J Clean Prod 226:1092–1101

Monroe MC, Plate RR, Oxarart A, Bowers A, Chaves WA (2019) Identifying effective climate change education strategies: a systematic review of the research. Environ Educ Res 25(6):791–812

Obasanjo O, Pinzó JC, Mills G, Hartley R, Hamukoma N, Calburn, Doyle N, Games D, Muzanda A, van der Merwe E, Kilcullen D, Davis D (2020) Future of African cities project. Special report—January 2020. Where the rubber hits the road. The Brenthurst Foundation, Johannesburg

Odendaal N (2012) Reality check: planning education in the African urban century. Cities 29(2012):174–182

Reid H, Sahlen L, Stage J, Macgregor J (2008) Climate change impacts on Namibia's natural resources and economy. Clim Policy 8:452–466

SACPLAN (South African Council for Planners (2014) Guidelines for competencies and standards for curricula development. Midrand. https://sacplan.org.za/wp-content/uploads/Competencies-Guidelines.pdf

Satterthwaite D, Archer D, Colenbrander S, Dodman D, Hardoy J, Mitlin D, Patel S (2020) Building resilience to climate change in informal settlements. One Earth 2(2):143–156

Scott D, Davies H, New M (eds) (2019) Mainstreaming climate change in urban development: lessons from Cape Town. UCT Press, Cape Town

Scott D, Iipinge KN, Mfune JKE, Muchadenyika D, Makuti OV, Ziervogel G (2018) The story of water in Windhoek: a narrative approach to interpreting a transdisciplinary process. Water 10(1366):1–16

Shaban ARA (2019) 90% Of Mozambican city of Beira destroyed by Cyclone Idai—Red Cross. Africa News, 18 March, 2019. Available at: https://www.africanews.com/2019/03/18/90-percent-of-mozambican-city-of-beira-destroyed-by-cyclone-idai-red-cross/. Accessed 21 July 2020
Stockholm Environment Institute (2018) From policy to practice—Annual report 2018. The Stockholm Environment Institute, Stiftelsen
Stone M (2020) A plague of locusts has descended on East Africa. Climate change may be to blame. National Geographic, 14 February 2020. Available at: https://www.nationalgeographic.com/science/2020/02/locust-plague-climate-science-east-africa/. Accessed 21 July 2020
Tervo-Kankare K, Saarinen J, Kimaro ME, Moswete NN (2018) Nature-based tourism operators' responses to changing environment and climate in Uis Namibia. African Geographical Review 37(3):273–282
UNESCO (United Nations Educational, Scientific and Cultural Organization) (2017) UNESCO at COP23. Climate change education. Available at: http://unesdoc.unesco.org/images/0026/002600/260083e.pdf. Accessed 22 August 2020
UNFCCC (United Nations Framework Convention on Climate Change) (2015) Paris Agreement. Available at: http://unfccc.int/paris_agreement/items/9485.php. Accessed 22 August 2020
Watson V, Odendaal N (2012) Changing planning education in Africa: the role of the Association of African Planning Schools. J Plan Educ Res 33(1):96–107
Van Rensburg P (2016) Overcoming global water reuse barriers: the Windhoek experience. Int J Water Resour Dev 32(4):622–636
Zellner M, Campbell SD (2015) Planning for deep-rooted problems: what can we learn from aligning complex systems and wicked problems? Plan Theory Pract 16(4):457–478. https://doi.org/10.1080/14649357.2015.1084360

Abraham R. Matamanda graduated from the University of Zimbabwe with a B.Sc. (Hons) in Rural and Urban Planning and an M.Sc. in Social Ecology. He holds a Ph.D. in Town and Regional Planning from the University of the Free State (UFS). Abraham lectures in the Department of Geography at UFS. His current research focuses on climate change adaptation, informal urbanism, medical geography and political economy. Abraham has presented his work at international conferences and contributed to national and international peer-reviewed journals. Among his recent publications is a book he co-edited: *Urban Geography in Postcolonial Zimbabwe: Paradigms and Perspectives for Sustainable Urban Planning and Governance*

Jennilee M. Kohima lectures in the Department of Architecture and Spatial Planning at Namibia University of Science and Technology. She obtained her M.Sc. in Land Management and Land Tenure from TU Munich in Germany and a Master of Urban and Regional Planning from the University of the Free State (UFS). She has over 10 years of work experience in the public, private, development and academic sectors in various development and planning fields. She is currently a Ph.D. candidate at UFS, and her research interests are around issues of urban informality, land use management, bottom-up planning and planning education

Verna Nel qualified as a town and regional planner at Wits University and obtained her M.Sc. and Ph.D. through University of South Africa. After three decades of working primarily for municipalities, she moved to the Urban and Regional Planning Department of the University of the Free State. She has diverse research interests that include spatial and urban resilience, local economic development and spatial governance. She has presented her work at international conferences and has published her research in leading journals and books

Innocent Chirisa is a Professor of Urban and Regional Planning in the Department of Demography Settlement and Development at the University of Zimbabwe. He is also a research fellow in the Department of Urban and Regional Planning at the University of the Free State in South

Africa. Currently, Prof. Chirisa is serving as Dean of the Faculty of Social and Behavioural Sciences at the University of Zimbabwe. His research interests are environmental systems, human settlements planning and resilience. To date, he has written 20 books, more than 80 book chapters and over 100 articles for peer-reviewed journals

Chapter 7
Is Climate Change Knowledge Making a Difference in Urban Planning and Practice: Perspectives from Practitioners and Policymakers in Tanzania

Wilbard Jackson Kombe and Samwel S. Alananga

Abstract The magnitude and effects of Climate Change (CC) such as floods and storms are projected to increase in the future. There is also a consensus among scholars that rich CC knowledge in urban planning can lead to better Climate Change Adaptation (CCA) and Mitigation (CCM) outcomes. However, generally the role of planners and plans in responding to Climate Change (CC) challenges has been disappointing and increasingly questioned. This chapter analyses the role of planning education, experience and/or practice among professional planners in addressing climate adaptation and mitigation issues. Field studies involving face to face interviews were conducted in Arusha Municipality in 2019. Questionnaires were completed by practitioners and policymakers. The findings highlight the gaps in CC knowledge and capacity among planners and policymakers. Also, the extent of informality, the major force transforming urban land use and development is overlooked. Most importantly, there is insensitivity, lack of accountability and political commitment by the Local Government Authority (LGA) on CC issues in planning, budgeting, and management. We argue that improving the role of urban planning in CCA and CCM requires: a recognition of the indispensable role of LGAs; substantive engagement of stakeholders; acknowledgement of socio-cultural and economic barriers to CCA/CCM at the local level; guidance on informality; and adaption of multi-level governance and integration of spatial and economic planning at city and community levels.

Keywords Adaptation · Mitigation · Planning · Climate change knowledge · Practice

---

W. J. Kombe (✉)
Institute of Human Settlement Studies, Ardhi University, Dar es Salaam, Tanzania
e-mail: wilbard.kombe@aru.ac.tz

S. S. Alananga
School of Earth Sciences, Real Estate Business and Informatics, Ardhi University, Dar es Salaam, Tanzania
e-mail: samwel.alananga@aru.ac.tz

G. Alem Gebregiorgis et al. (eds.), *Planning Cities in Africa*, The Urban Book Series,
https://doi.org/10.1007/978-3-031-06550-7_7

7.1 Introduction

Climate Change (CC) impacts present challenges to cities and municipalities around the globe (Kiunsi 2013). Herslund et al. (2015) and Roberts (2010) note a limited momentum among stakeholders for a broad citywide Climate Change Adaptaion (CCA) strategy to address the complexity of CC impacts due to capacity constraints. The city's poor are currently unable to cope adequately with the existing climate variability while the rapid urban growth compounds the current situation (Pan African, START, Tanzania Meteorological Agency, and Ardhi University 2011). From a spatial planning perspective, it is the role of urban planners to coordinate, for example, local preferences/contexts and stakeholders' initiatives with the (sectoral) policy objectives, from the local to the (inter)national level (Biesbroek et al. 2009).

CC response requires reshaping of local planning, regulatory and financial frameworks to encourage and support adaptation of households, community organisations, NGOs and the private sector, and local government responses for risk reduction (Moser and Satterthwaite 2008). Since Climate Change Mitigation (CCM) strategies may have prospects of effectiveness only after a very long period of time, it may result in less CCA, however, for spatial planning that is of limited relevance (Biesbroek et al. 2009). It is anticipated that through spatial planning, Local Government Authorities (LGAs) can increase resiliency to major climate shifts and ensure communities are equipped with built-in mechanisms to adapt and mitigate such changes (Hirokawa and Rosenbloom 2013).

Furthermore, the extent of municipal Climate Change Policy (CCP) tends to increase with the attainment of education and awareness among local residents as has been investigated in previous studies (Vasi 2006; Feiock et al. 2009; Krause 2010, 2013; Pitt 2010; Sharp et al. 2011; Kwon et al. 2014; Opp et al. 2014). Low levels of education and commitment among elected officials in many developing countries compound the challenges of CCP. There is, therefore, a pressing need to examine the extent of CC knowledge and awareness among professionals involved in planning at the local level. However, the level of knowledge on CC impacts, adaptation, and mitigation strategies among urban planners in developing countries has not been adequately investigated. This chapter draws from the experiences of spatial planners and policymakers in Tanzania, with respect to both knowledge and understanding of CC impacts, adaptation and mitigation in relation to CC knowledge, as applied to diverse plans and related spatial tools.

7.2 Conceptual Framework

The term governance refers to the different ways in which societies can organise themselves and make decisions to achieve Climate Change Adaptation (CCA) or Climate Change Mitigation (CCM). Resilience is understood as an attempt to return to normal with normality being the pre-disaster state (Kelman et al. 2015). Resilience

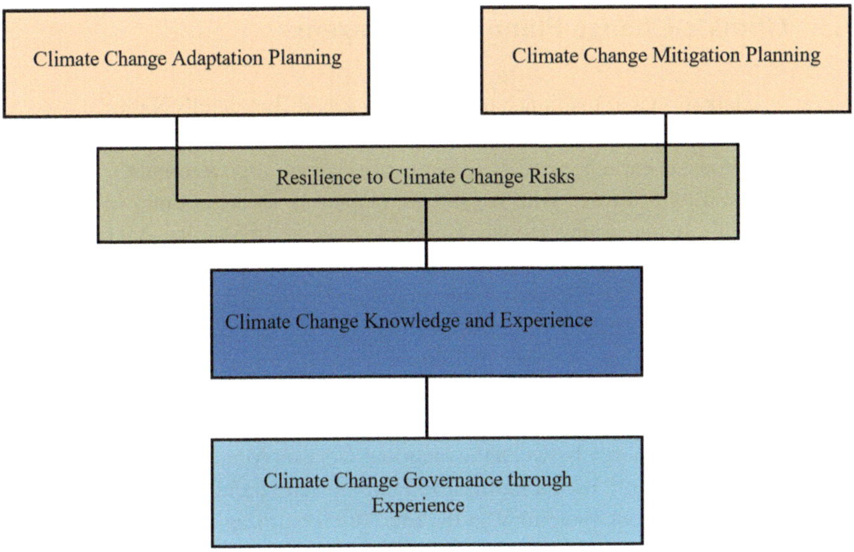

Fig. 7.1 Climate change responsive planning (*Source* Authors' 2021)

can also be considered as the ability of a system to withstand a major shock and maintain or quickly return to normal function (Leichenko 2011). Building a city's resilience requires a multi-level, cross-sectoral, as well as multi-stakeholders' governance approach (Dodman et al. 2011). Assessing the level of the urban planners' knowledge of CCP provides a mechanism to assess the importance of urban planning training programmes as summarised in Fig. 7.1.

In the Global South, CCP is not well enshrined in their public works programmes at the neighbourhood level (Storch et al. 2011). In addition to the demand for professional, urban planning education, most developing countries are facing several other challenges including the weak capacity of LGAs, short-term planning, and physically oriented plans that do not respond effectively to change (Roberts 2010; Herslund et al. 2015). Furthermore, many plans end up not being implemented (Storch et al. 2011). A further challenge to CC responsive planning is that LGAs often lack the commitment to environmental planning and the fact that higher authorities tend to impose the requirement to engage in CCP activities in a top-down approach (Burby 1998). The extent to which municipalities engage in CCP tends to increase with the commitment levels of elected officials, planning staff and local residents (Bedsworth and Hanak 2013; Stevens and Senbel 2017; Shi et al. 2015).

7.3 Climate Change Planning in Tanzania

As far as adaptation and resilience to CC is concerned, Tanzania's National Adaptation Programme of Action (NAPA) and National Climate Change Strategy (NCCS) provide the most comprehensive statements of the central government's framing of adaptation and its priorities with respect to adaptation as the country lacks a standalone climate change policy. However, Smucker et al. (2015) argue that CCA policy constructs an anti-politics of adaptation through its framing of CC as an urgent and generalised threat to development while failing to sufficiently address the complex governance and social equity dimensions of CCA.

Kiunsi (2013) argues that the national CC strategy has limited provision for measures to build adaptive capacity in Tanzania. Malele (2017) has forwarded a further argument identifying a weak policy framework for disaster risk management. He notes a lack of synergy between the National Disaster Management Policy (URT, 2007) and sectoral policies including the Human Settlement Development Policy (2000). There is also an agreement in the literature regarding the existence of loopholes in the current policies and regulations. For instance, with respect to ownership, use, and management of the hazardous lands and protected areas are regarded as "no one's land" and therefore act as a setback towards Disaster Risk Reduction (DRR) as well as to building a city's resilience (Kweka 2017; Malele 2017).

Additional challenges to CCP in Tanzania are: hierarchal government structure in which adaptation and disaster risk management are centralised in the National President's and the Prime Minister's Offices respectively, with no staff and resources at LGAs levels (Vedeld et al. 2015; Kweka 2017); a multi-level and fragmented governance system, which limits an integrated approach to building resilience (Herslund et al. 2015; Vedeld et al. 2015; Malele 2017); weak enforcement of existing regulations, especially those related to urban planning and environment management (Malele 2017; Kweka 2017; Sakijege et al. 2012; Rugai and Kassenga 2014); and inadequate implementation of policy provisions to enhance resilience in urban areas, e.g. the provisions needed to provide infrastructure and services prior to allocating plots to developers (URT 2000; Malele 2017).

At the household level adaptation to CC related challenges has been thoroughly studied (John et al. 2014; Kiunsi 2013; Sakijege et al. 2012), but CCP and risk minimisation measures through structural adaptation strategies have hardly been achieved at all at the household level. Kiunsi (2013), Parnell (2015) and Sakijege, Lupala and Sheuya (2012) argue that these adaptation strategies are constructed with little or no attention to acceptable technical considerations and with limited capacity. The issue of lack of capacity in relation to planning and implementation of CCA measures is broadly discussed in the literature on Tanzania (Dodman et al. 2011; Rugai and Kassenga 2014; Shemdoe et al. 2015). The deficit of budget in LGAs compounds the challenges related to building resilience in favour of either CCA or CCM (Kithiia 2010; Kiunsi 2013; Malele 2017). The problem of enforcing land use zoning is compounded by other factors, including interference from leaders for political reasons, as well as malpractice including corruption (Vedeld et al. 2015; Rugai and Kassenga 2014; Kweka 2017; Shemsanga et al. 2010).

Arguments on improving adaptation measures and capacities in the literature include social changes for adaptation, primarily related to education and capacity building; scientific change in terms of detailing the implications of CC, including detailed cost estimates of temperature change, sea-level rise and changes in precipitation (Dodman et al. 2011). Carmin et al. (2011); and a call for the need to link adaptation planning and implementation to city priorities and existing initiatives, as CC frequently is perceived as being a distant threat, and therefore, regarded as less important and often in conflict with immediate priorities.

Kithiia (2010) argues that city planners and decision-makers view CC through the lens of the global policy agenda, hence there is no CC response planning and/or failure to provide for CCA planning within the overall municipal development planning. In Dar es Salaam critical risk reduction infrastructure such as stormwater drainage, sanitation systems, and solid waste management is inadequately provided for in the settlement plans (Malele 2017). At the same time, urban planning is often done with limited involvement of stakeholders at various levels (Vedeld et al. 2015). Kweka (2017) argues that the state controls land use planning in a hierarchical and non-transparent manner despite the national policies and acts calling for transparency, accountability, and participation of other key stakeholders. Since most of the CCP techniques are not well implemented in Tanzania, the literature emphasises the need for capacity building of the technical cadres (Shemdoe et al. 2015).

7.4 Research Methodology

This study links knowledge of CC among professional planners in Tanzania aiming to understand how this has impacted resilient cities and the practices on the ground. This is achieved through responses to the following research questions:

- How resilient are local authorities in Tanzania, considering the degree of adoption of either CCM or CCA planning?
- To what extent is the adoption of CCM or CCA planning related to planners' knowledge gained through multiple sources?
- What are the variations in both the adoption of CCM and CCA planning and the level of CCP knowledge among urban planners in Tanzania?

Two approaches are implicit in this study; the first is a questionnaire survey which was conducted in Arusha. The Arusha Master Plan was explored in this study by interviewing planners who were working for Arusha District Council (ADC), Arusha City Council (ACC) and Arumeru District Council (ArDC) during the period of the Master Plan preparation. The survey employed a semi-structured questionnaire. The questionnaire was designed to capture a number of specific experiences among planners such as the awareness and education variables, whereby in the first part of the questionnaire the awareness and CC education indicators were combined with the experience indicators to generate an aggregate indicator of knowledge on CCP[1]; then the level of implementation of plans in terms of eight adaptation planning variables and 13 mitigation planning variables were combined to construct an aggregate indicator of CCP[2]; the last section captured the observed climate abnormalities and the associated impacts. The survey responses were analysed based on the knowledge and planning indices coupled with cross tabulations which were then combined to generate information rich tables for interpretation.

For purposes of triangulating the results, it was necessary to include several interviews with experienced senior planners from one of the LGAs that has recently prepared and is implementing a land use master plan. The preparation of the Arusha Master Plan (AMP) was an initiative of the Ministry of Lands, Housing and Human Settlements Development (MLHHSD) to develop a comprehensive plan for the city. The 2014 AMP was intended to take stock of past planning and urban development initiatives in Arusha and develop a detailed master plan for the city and its catchment areas, such as peri-urban villages of Arusha and Arumeru Districts. To probe the details of the plan, discussions were held with retired planners who worked in Arusha for more than ten years and senior planners who were currently working. In addition, local leaders in the study area and members of the community were interviewed. In 2019, local community leaders facilitated the process of searching for people and the arrangement of interviews. The analysis of interviews was through discretizing of narratives so as to organise them into specific areas and with issues that emerged in the questionnaire.

7.5 Results and Discussion

7.5.1 Respondents' Profile

Table 7.1 suggests that approximately 46% of planners reached in this study have not had any formal training on CC. In terms of age, Table 7.1 suggests that the likelihood of attending a training programme on CC could be the same regardless of age; the younger planners seem to have a greater likelihood of not having attended any such CC training. Furthermore, the findings indicate that the likelihood of attending CC training among planners is relatively higher among females than males. Also, around 30% of those with the longest workplace stay attended CC training while none of those with the shortest workplace stay (a year or less) have ever attended

[1] CCP knowledge is an aggregate indicator of the probability of having adequate climate change knowledge based on experience, duration of stay at a workplace, education, knowledge of policy, and attendance at formal training in climate change. It is computed by totaling the weighted scores for each of the items and dividing by the total possible weighted score. The weights are the likert scale measures at a distance of 1.

[2] This was constructed in a similar manner as CCP knowledge using indicators for CCP and adaptation planning.

Table 7.1 Description of survey respondents in relation to training on climate change

	Category	Attended training related to climate change		Total
		No	Yes	
Age	Above 40	20	9	29
	40 and below	26	9	35
Total		46	18	64
Gender is male	No	8	5	13
	Yes	38	13	51
Total		46	18	64
Workplace stay	1 year or less	2	0	2
	2–3 years	9	5	14
	4–6 years	6	1	7
	7–9 years	1	2	3
	Over 10 years	7	3	10
Total		25	11	36
Higher Education Degree or above	No	3	1	4
	Yes	43	17	60
Total		46	18	64
Practising as planners	No	17	3	20
	Yes	29	15	44
Total		46	18	64
Experience category	3 years or less	7	1	8
	4–6 years	8	3	11
	7–9 years	2	2	4
	Over 10 years	27	10	37
Total		44	16	60

Source Authors' Fieldwork 2019

similar training. Formal education for planners offers an opportunity to acquire new knowledge, especially on CC (Vasi 2006; Krause 2013; Pitt 2010; Opp et al. 2014), which is, however, dependent on the trainers' curriculum. Lastly, it is notable that with relatively more experience come extra opportunities to access knowledge on CC.

7.5.2 The Need for Local Context and Stakeholders' Engagement

The need for local context and stakeholders' engagement is well enshrined into planning policies in Tanzania (URT 2000), though the extent to which they are implemented varies considerably. The preparation of the AMP was contracted to an international consulting firm presumably with planning and CC knowledgeable staff. Also, key stakeholders were involved including councillors from the ACC and ArDC. These stakeholders participated in consultation meetings and workshops. In addition, a town planner from each of the three LGAs of ACC, ADC and ArDC was appointed to work full-time with the consultants as part of the local capacity building. In this regard, the consultant did receive local support and had contacts who were required to help them gain a good understanding of the local contexts. This was confirmed by one of the respondents from the local community:

> … we are aware of the AMP, we participated in the meetings held at the ADC. I do not think there are many people who are ignorant about the AMP. There was a lot of publicity and community leaders, politicians and media were involved. ….

Although the role of participatory planning is well articulated in the literature (Vedeld et al. 2015; Kweka 2017), its ability to capture local contexts in spatial planning is context-specific. The interviews repeatedly revealed that community leaders and representatives of residents in Sokoni II (an area within the AMP) and other areas were engaged during the preparation of the AMP, and yet their concerns were not addressed in the AMP. This is reflected in the following excerpt:

> …, we proposed that we want this area to remain our residential area, but they have proposed high income housing. This is not only a problem in Sokoni II, but in other peri-urban settlements governed by the ACC also. For instance, in Kijenge they have earmarked the area already occupied by residents as a sports complex area.

Similarly, defining the limits of the AMP was also problematic ab initio. This was because preferences of the sitting land occupiers to be part or not part of the city were ignored, a situation that is well illustrated in the literature (Moser and Satterthwaite 2008; Bedsworth and Hanak 2013; Biesbroek et al. 2009). The city planner interviewed noted:

> …many landowners in Sokoni II and other peri-urban villages under the ACC do not want to be re-categorised as urban residents. They prefer to remain rural because this gives them opportunities such as the freedom to use or build on their land or subdivide and sell/transfer without land use planning. They are also not obliged to pay land taxes.

In this regard, implementation of a plan may not only be facilitated by community participation, depending on education levels, it may be hampered by it as well (Feiock et al. 2009; Krause 2013; Pitt 2010; Sharp et al. 2011; Opp et al. 2014). What is important is that people on the ground must understand what the pros and cons are behind any proposed plan. Preferences and wishes might significantly differ between different stakeholders, but ultimately, they must be harmonised in order to arrive at viable plans (Bedsworth and Hanak 2013). Local context needs to be captured as people in developing countries have a strong connection to their land as asserted by the following respondents:

> …our land is everything. We depend on it for food and income – most of the fresh vegetables, fruit, milk and so on in the city are from this area and other peri-urban settlements of the ACC. Most sitting land occupiers prefer to remain rural so as to maintain their rights and interests; but this is not what the ACC would wish.

> … land especially within the peri-urban villages of the ACC is a contentious issue. It is very difficult for people in the peri-urban areas to understand why one would be part of the ACC. People are also reluctant to contribute land for public uses although they are aware of the master plan and also want better infrastructure services. We have faced this problem in the projects to regularise informal settlements. It remains unresolved despite the existence of a master plan.

Elaborating on the unregulated land use changes and development in the peri-urban villages under the ACC, a respondent (LGA senior staff) noted:

> …ACC and ADC have had conflicting interests over the peri-urban land which surrounds the ACC which is administratively and legally an entity of the ADC (a rural district authority), but geographically and functionally more or less an integral part of the ACC. The ACC wants this area to be declared urban i.e., part of the ACC, whereas the ADC maintains that this is a rural area which ought to be under the ADC. None of the senior officials including the councillors in the two LGAs seem concerned about the increasing climate change induced threats such as flooding and landslides, which start in the upstream settlements (ADC) and extend downstream (ACC) and would require an integrated CCP and implementation. The dispute between the ACC and the ADC is not really about land but mainly about revenue generation activities which each of the LGAs want to access.

This excerpt suggests that even with good CC knowledge, the existing institutional setting including the laws and vested interests of policymakers, can be a real challenge to addressing and implementing CC responsive planning (Moser and Ekstrom 2010; Herslund et al. 2015).

Regarding the weak consideration of the grassroot preferences the city planner interviewed noted:

> …conceptual ideas and sketches were developed by the team and local actors in Arusha; but a team which was based in Singapore generated the master plan; this was a detailed master plan with illustrations. Many good ideas were lost at the later stage because of little or lack of reference to the real conditions on the ground. The concept of participation seems to have been taken for granted and this entailed no involvement, i.e. no wonder many ideas, were collected and deliberated upon by stakeholders but not incorporated into the master plan.

The consultants for the AMP seem to have fallen into the classical trap of flying too high, failing to develop deeper insight and understanding of the contextual conditions that have for decades driven the growth of Arusha, as well as significantly shaped its current urban form and structure (Vedeld et al. 2015; Kithiia 2010; Kweka 2017). This refers, for instance, to the existing customary and quasi-customary land tenure rights and practices, and unregulated land use changes which are propelled by informal land parcelling, transactions, and development in the peri-urban areas of Arusha City including Sokoni II. Like many other major cities in the Global South, informality is the single most critical factor that is transforming Arusha and other large cities in Tanzania (Kombe 2017).

7.5.3 *Planners' Knowledge and Observations of Climate Anomalies*

The major challenge that any planning consultant would face in Tanzania is to come to an appropriate understanding of how much knowledge there is on the ground. In the case of CC, for example, Table 7.2 presents CC-related anomalies based on planners' experience in their respective regions. The idea is that in addition to formal training, local climate knowledge can also be acquired through experiencing the changing climate-related impacts. The figures in Table 7.2 suggest that the Coast region has experienced an upsurge in climate anomalies while responses from Dar es Salaam differed slightly in all of the three items evaluated. However, the responses suggested an 81% probability of an upsurge in climate anomalies in Dar es Salaam. From the responses, the only regions that suggested a slight downward spiral in climate anomalies were Dar es Salaam, Singida, Dodoma and Tabora, with Tabora suggesting the greatest downward spiral in climate anomalies of about 33%. Generally, planners

Table 7.2 Stated climate anomalies across regions

		Temperature anomalies		Precipitation anomalies		Sea level anomalies		Overall	
		Down	Up	Down	Up	Down	Up	Down	Up
Workplace region	Coast	0.00	1.00	0.00	1.00	0.00	1.00	0.00	1.00
	Dar es Salaam	0.08	0.85	0.08	0.85	0.00	0.73	0.05	0.81
	Iringa	0.00	1.00	0.00	1.00			0.00	0.67
	Mtwara	0.00	0.67	0.00	0.75	0.00	0.25	0.00	0.56
	Simiyu	0.00	0.50	0.00	1.00			0.00	0.50
	Mwanza	0.00	0.33	0.00	1.00			0.00	0.44
	Singida	0.33	0.67	0.00	0.67			0.11	0.44
	Dodoma	0.14	0.59	0.23	0.45			0.12	0.35
	Arusha	0.00	0.00	0.00	1.00			0.00	0.33
	Tabora	0.00	1.00	1.00	0.00			0.33	0.33
	Rukwa	0.00	0.00	0.00	0.00			0.00	0.00
	Not Mentioned	0.00	0.88	0.13	0.75	0.14	0.57	0.09	0.73
Overall		0.08	0.68	0.15	0.66	0.20	0.54	0.14	0.63

Source Authors' Fieldwork 2019

anticipate upward climate anomalies in the four regions of Coast, Dar es Salaam, Iringa, Mtwara and Simiyu. The remaining regions are predominantly normal, based on the opinion of planners.

Responses from Arusha suggest a 33% chance of an upsurge in CC anomalies largely due to a 100% upsurge in precipitation. If this information alone had been captured in the AMP, the consultants would have taken due care to question local communities in flood-prone areas. However, the ideas that seem to frame the concept and structure of the Arusha Master Plan (2035) and its major components have been largely drawn from general information and data on environment, social, economic and institutional contexts and some imported concepts. For instance, the proposed new Arusha Central Business District (CBD) draws inspiration from the Singapore CBD, whereas the proposed residential development in the CBD seems to have been animated by London neighbourhood redevelopment. In addition, the new fringe centre and business park and the proposed business aviation and medical tourism centre have been inspired by the Balliol Business Park Wolverhampton[3] and the Saint Tropez Airport[4] France, respectively. The AMP provides strong evidence of a case where planners have uncritically adapted from borrowed urbanism (Ong 2011). This approach has repeatedly failed to deliver appropriate results in most cities in the Global South. Although knowledgeable people were involved in the preparation of the AMP, there is little to suggest that their knowledge was contextualised to capture local experiences.

The observations on the AMP processes provide an opportunity to further re-examine the type of knowledge held by planners, specifically on which aspects of human settlements are extremely affected by CC. Table 7.3 analyses the responses in terms of six potential areas of CC impacts. From a planners' perspective, human settlement is the most affected sector, potentially reflecting upon their sphere of knowledge. This is followed by water resources and ecosystem changes. Region-wise the greatest, overall worsening impacts are observed in Mbeya, Pwani, Simuyu and Singida. However, in terms of human settlements, the greatest worsening impacts of CC were reported by planners from Arusha, Mbeya, Mtwara, Pwani, Simuyu and Singida. This is, however, the case if the responses are not weighted by the number of responses. With respondents factored in, Dar es Salaam has the worst of all CC impacts. Planners in Arusha suggest that Arusha is greatly affected by CC impacts, especially to human settlements, a view which should have been incorporated into the AMP.

[3] Wolverhampton is a town in the Midlands, north of Derby and Birmingham.

[4] Saint Tropez Airport is an airport located in La Môle, 15 km (8 NM) southwest of Saint-Tropez, in the Var department of the Provence-Alpes-Côte d'Azur region in southeastern.

Table 7.3 Planners' predicted probability of worsening climate-induced changes in their workplace regions in Tanzania

	Arusha	Dar es Salaam	Dodoma	Iringa	Mbeya	Mtwara	Mwanza	Pwani	Rukwa	Simiyu	Singida	Tabora	Not mentioned	Overall
human settlement changes	1.00	0.86	0.65		1.00	0.33	1.00	1.00	0.00	1.00	1.00	0.00	0.63	0.72
water resource change	0.00	0.86	0.59	0.00	1.00	0.67	1.00	1.00	0.00	1.00	1.00	0.00	0.75	0.70
ecosystem changes	0.00	0.79	0.56	0.00	1.00	0.33	0.67	1.00	0.00	1.00	1.00	1.00	0.75	0.67
human health change	0.00	0.71	0.60	1.00	1.00	0.50	1.00	1.00	0.00	1.00	1.00	0.00	0.50	0.66
food insecurity	0.00	0.92	0.42	0.00	1.00	0.50	0.67	1.00	0.00	1.00	1.00	0.00	0.71	0.63
social life change	0.00	0.67	0.50	1.00	1.00	0.33	0.67	1.00	0.00	1.00	1.00	0.00	0.57	0.58
Overall	0.33	0.83	0.60	0.00	1.00	0.44	0.89	1.00	0.00	1.00	1.00	0.33	0.71	0.70

Source Authors' Fieldwork 2019

7.5.4 *Planners' Climate Change Knowledge and Planning Practices*

To obtain some insights into CCA strategies across the urban centres which the respondents represented, Table 7.4 provides a summary of the observations on the implementation of different CCA strategies. Across regions, earmarking and implementation is only relevant for "raising public awareness and participation" and for the other strategies the planners suggested implementing without earmarking. Based on current planning practices it seems adaptation measures have limited scope in the planning process, partly because adaptation is principally carried out on an individual basis in reaction to a perceived risk at a particular point in time. This however does not mean that planning is irrelevant, but that the inefficiency of individual actions to address a spatially spreading problem requires coordination in planning (Herslund et al. 2015).

Contrary to CCA strategies, most CCM strategies require adequate spatial planning prior to implementation. The AMP considered and provided for passive and active areas for recreation; agricultural land; water bodies; nature conservation; parks

Table 7.4 Planners' ranking of adaptation planning in their respective LGAs

S/N	Adaptation planning strategies	NA	1	2	3	4	TWS	% Score
1	Has raising public awareness and participation been earmarked for implementation in your LGA?	8	2	3	4	22	108	0.87
2	Has extension of basic urban services and control of urban growth boundaries been earmarked for implementation in your LGA?	9	6	6	2	16	88	0.73
3	Has implementing specific programmes for waste and stormwater management been earmarked for implementation in your LGA?	8	7	5	6	13	87	0.70
4	Has encouraging water sensitive land use (agriculture/industry) been earmarked for implementation in your LGA?	10	8	7	5	9	73	0.63
5	Have activities that concern the adoption of inter-organisational coordination procedures been earmarked for implementation in your LGA?	10	10	7	5	7	67	0.58
6	Has carrying out risk assessment been earmarked for implementation in your LGA?	9	11	9	1	9	68	0.57
7	Has low impact design for impervious surface been earmarked for implementation in your LGA?	11	11	9	4	4	57	0.51
8	Has adopting of disaster-resistant land use and building codes been earmarked for implementation in your LGA?	9	15	7	1	7	60	0.50

NB: 1 = Neither earmarked nor implemented; 2 = Earmarked not implemented: 3 = Implemented without earmarking, and 4 = Earmarked & implemented
Source Authors' Fieldwork, 2019

and so on. Also, innovative ideas on public transport, such as a three corridor rapid bus transport (BRT) system were proposed. These described provisions constitute CCM strategies. which are better placed for planning than CCA. Table 7.5 provides the

Table 7.5 Ranking of mitigation planning in LGAs

S/N	Mitigation planning	NA	1	2	3	4	TWS	% Score
1	Has infill development & reuse of brownfield sites been earmarked for implementation in your LGA?	12	16	5	6	39	200	0.76
2	Has raising public awareness and participation been earmarked for implementation in your LGA?	11	4	7	5	12	81	0.72
3	Has vegetation (forests/woodlands) protection been earmarked for implementation in your LGA?	8	6	8	5	12	85	0.69
4	Has creation of conservation zone protection areas been earmarked for implementation in your LGA?	12	6	8	2	11	72	0.67
5	Has mixed use or compact development been earmarked for implementation in your LGA?	12	8	6	4	9	68	0.63
6	Has watershed-based/ecosystem-based land management been earmarked for implementation in your LGA?	12	7	9	4	7	65	0.60
7	Has green building/infrastructure been earmarked for implementation in your LGA?	10	11	5	5	8	68	0.59
8	Has multi-modal transportation corridor improvement been earmarked for implementation in your LGA?	12	10	7	5	5	59	0.55
9	Has pedestrian/resident-friendly, bicycle-friendly, transit-oriented community design been earmarked for implementation in your LGA?	11	10	9	3	6	61	0.54
10	Has adopting inter-organisational coordination procedures been earmarked for implementation in your LGA?	12	10	11	4	2	52	0.48
11	Has energy-efficient land use been earmarked for implementation in your LGA?	13	13	8	3	2	46	0.44
12	Has zero waste recycling been earmarked for implementation in your LGA?	12	16	5	3	3	47	0.44
13	Has adjusting by-laws for vehicle emission reduction been earmarked for implementation in your LGA?	12	22	2	1	2	37	0.34

NB: 1 = Neither earmarked nor implemented; 2 = Earmarked not implemented: 3 = Implemented without earmarking and 4 = Earmarked & implemented
Source Authors' Fieldwork, 2019

ranking of earmarking and implementation of 13 CCM strategies by respondent planners. Table 7.5 shows that only "infill development & reuse of brownfield sites" were earmarked and implemented beyond 50% (these score 76%). Other strategies scored above 50% only on one side of earmarking or implementation: they include "raising public awareness and participation", protection of "vegetation (forests/woodlands)"; "watershed-based/ ecosystem-based land management" and "creation of conservation protection zones or areas". These percentages suggest that many of the CC strategies, for example, conservation areas, can be implemented in many of the cities in the Global South even if not formally declared as conservation zones. For instance, in many traditional societies of Tanzania, conservation of nature has been among the prime objectives of customary land use controls. This was also evident in the case of the AMP where the interviews of the Project Affected People (PAP) suggested a strong tradition of environmental conservation strategies. This is evident in the following excerpt from one interviewee:

> ...regarding the protection of nature (i.e., forest, streams, and soil) people are aware, although they might not say so, that most landowners grew up in this area (Sokon) when it was a vast expanse of evergreen. Protection of the environment is part of our culture and was held in high regard to our parents and grandparents. We have local norms, values, and traditions to protect the natural environment, such as forests, water sources and streams, and valleys. In the 70s and 80s the government conducted training on environmental and soil conservation (terracing) and protection of traditional exotic tree species (such as Pinus patulaand, Eastern red cedar). People were forbidden from cutting down such trees and other plants in the forest. We were told frightening stories about the existence of wild man-eating animals in the public forest bordering the settlement (Sokoni); these were intended to scare residents from encroaching on the public forest. The situation has changed now, the streams, valleys, and trees in Sokoni II are disappearing.

The potential for CCM strategies within the local customary environment is also reflected in the following excerpts:

> ... in the old days, we were only allowed to cultivate the land once a year in order to protect the environment. Also, terracing was mandatory, agricultural officers used to visit and inspect farms in Sokoni II and other areas; those who were doing very well were rewarded...

The preceding excerpt underscores the carrot–stick approaches in the traditional agricultural practices aimed to conserve the environment. The extensive conservation of green and blue areas constitutes an important consideration towards the amelioration of adverse impacts of CC and variation. Yet it is difficult to imagine how such ideas would be achieved given the unresolved and contentious customary land tenure issues involving vested interests in the peripheries of Arusha, as well as the Sokoni II settlement. There are also challenges in regulating in use and development in areas under customary land tenure as illustrated in the following excerpts:

> ...given the high demand for vegetables and fruit in the city, the lack of employment or alternative income generation opportunities coupled with the high price of land, it is difficult to convince farmers to abide by the traditional farming practices. Nowadays, cultivation is done throughout the year; no wonder erosion and landslides are increasing.

Population pressures, and communication and technological changes are basically alien to customs and traditions in agriculture. These coupled with the monetisation of rural economies, are exerting pressure on the limited land resources. The challenges facing the conservation of land under customary tenure are further illustrated as follows:

> …rapid population growth and socioeconomic changes in the city have pushed the demand and price of land up. This is especially so in settlements such as Sokoni II, Kwa Mrombo, Sakina, Mbauda, and Sekei. Most importantly, traditional customs require that a head of household parcels and allocates housing land to his grown-up sons. My father had five acres, I inherited only two acres in the 1970s, my brother got the rest. About ten years ago, I gave each of my four sons a place to build their own house, i.e., less than half an acre. I also sold a small portion to a relative. I am now only left with a small piece for my last born. Because of the shrinking size of land holdings, increasing demand and rising prices, it is difficult to protect fragile areas, especially those within the residential and farming areas.

Traditional conservation strategies, such as protection of plants on steep slopes and valley terracing, and fallow cultivation, have failed largely because of increased demands for land, and unregulated market forces that have exerted pressure on land under customary tenure. Most individuals are also concerned about the use of their land parcel without any coordination with adjoining landholders. Despite its significance, especially in food production, Arusha District Council officials are seldom involved in monitoring conservation matters in the peri-urban areas. Current land use practices are increasing runoff upstream and flooding in the downstream areas of Arusha. The interviews revealed that floods have increased because of human activities such as the clearing of vegetation and house construction.

> …floods have frequently affected many areas of the city, this includes the 2017 event which led to the loss of life and extensive damage to roads, drainage systems and housing in the midstream and downstream areas such as the Phillips, Njiro and Kwa Mrombo settlements. The loss of property and life is a result of the extensive clearance of vegetation, unregulated land use changes, and construction activities upstream, as well as encroachment on the public forest, steep hill slides, and streams and rivers in the peri-urban villages.

LGAs may earmark the four CCM strategies in Table 7.5 (S/N 1–4) in their plans, but implementation is likely to raise many challenges primarily because of the low priority given to CC matters by bureaucrats and little knowledge on the subject among decision-makers; also, weak institutional coordination within the City Council hampered by the compartmentalised departments of the Council. Most importantly, the fact that multiple authorities are involved in the land earmarked for conservation in urban plans, and weak or lack of vertical coordination are also threats to the implementation of CCM and CCA strategies (Herslund et al. 2015). There are also challenges associated with the malpractices in the informal land development and environmental conservation practices. Mistrust is evident among individuals towards community leaders primarily because many are reported to be involved in corruption in the informal land transactions (Kombe 2017). The following excerpt further illustrates:

> … when people subdivide, sell, or build with no regard for the master plan/building regulations, we cannot do anything. I once asked my neighbour who had bought land and started building a house very close to a stream, why he was doing so. He angrily responded that I have hatred and I am jealous because he (the neighbour) bought the land without involving

me. I felt bad; this is not unusual. Often local leaders do not intervene because they are accomplices in land deals, they get a commission. In the past elderly, traditional leaders were highly respected but these days they are also ignored.

The proceeding excerpt signals mistrust, and lack of commitment among elected leaders which hampers CCP in developing countries and particularly in Tanzania (Stevens and Senbel 2017; Shi et al. 2015; Bedsworth and Hanak 2013).

Formal training on CC, professional experience, and tacit traditional knowledge on customary practices in conservation constitute important sources of knowledge on CCP. Professionals are expected not only to involve local communities but also incorporate their proposals and experiences which have proved useful in planning. The relationship between CCP and knowledge for planners working in Dar es Salaam is presented in Fig. 7.2 while the same relationship for regions other than Dar es Salaam, Dodoma and Mtwara is presented in Fig. 7.3. It is clear from both figures that CC knowledge enhances CCP, with the exception of very high levels of knowledge for which the observations are contradictory. These contra views reflect highly knowledgeable planners who are unable to apply their knowledge because of either environment, regulatory constraints, or limited political commitment as previously noted. For instance, any attempt to accommodate all the key knowledge issues (i.e.

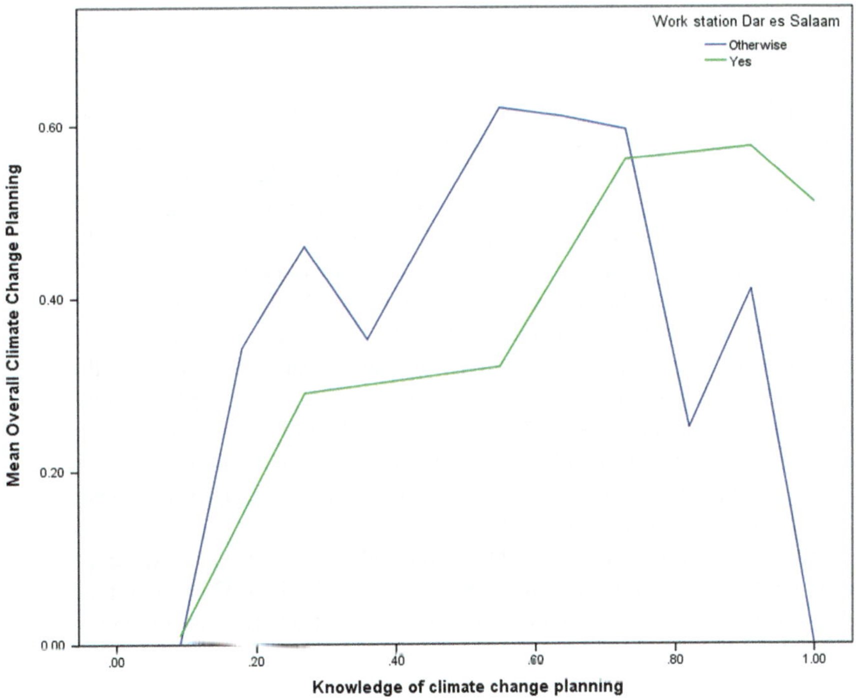

Fig. 7.2 Proportion of climate change planning in relation to knowledge (*Source* Author's Fieldwork 2019)

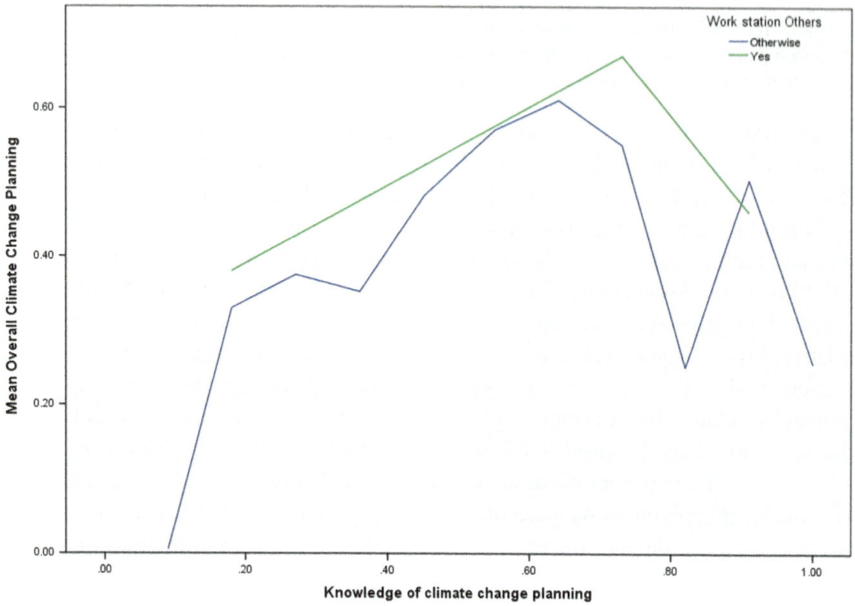

Fig. 7.3 Climate change planning and knowledge for regions other than Dar es Salaam, Dodoma and Mtwara (*Source* Author's Fieldwork 2019)

some of the ideas presented in Table. 7.4 and 7.5), covering formal, and customary knowledge and practices into a plan, may cause the plan to be rejected or delay its approval due to conflicting or vested interests of particular business elites.

Assessment of Fig. 7.2 suggests that Dar es Salaam, Dodoma and Mtwara planning practices are weak in terms of CCP. This could result in the unregulated land development or informality which predominates not only in Dar es Salaam but also other major cities of the Global South (Kombe 2017). Respondents also confessed that current land market dynamics, especially high prices, were a real challenge to the conservation agenda and initiatives. Just as in Dar es Salaam and other urban centres Informality does arise as a serious problem for the AMP (Kombe 2017). Experience of planners from Dar es Salaam and similar cities are seldom shaped by conservation practices of local communities since environmental conservation of forests (except large protected public forests outside the city), rivers, watershed and the like is seldom observable in the major cities.

The observation in Dar es Salaam could also be relevant in Arusha, but data limitation prevented a similar analysis as in Fig. 7.2. The AMP proposal to resettle peri-urban land occupiers in areas such as Sokoni II ignored the socio-cultural and economic needs and interests of sitting land occupiers. Reacting to this a respondent furiously noted:

> … instead of supporting livelihoods and basic services such as storm water drains to control erosion, landslides upstream, and floods downstream the AMP proposed to evict and resettle

us. How do they expect us to give away our ancestral land which is a walking distance to the city centre?

Any meaningful engagement with stakeholders, particularly local communities and institutions at Mtaa, Ward levels, and LGAs, ought to inform and generate knowledge, on the basis of which context for planning for CCA and CCM has to be grounded, although corruption in the land sector and lack of political commitment are real barriers to such initiatives.

7.6 Conclusion and Recommendation

Based on the observations in this study, a broader definition of CC knowledge inevitably entails an integrated approach that creates a positive relationship between CC knowledge and CCP. The challenge is, however, to capture local knowledge and experiences for incorporation into spatial plans. There are also disparities in the relationship between CC knowledge and CCP for major cities as contrasted to smaller cities. Larger cities are less likely to capture local conservation strategies as part of CC initiatives, thus lagging behind rural and smaller towns in CCP endeavours. With these observations being triangulated in the case of the AMP in terms of the concerns at the local level coupled with several other challenges, it is suggested that planning knowledge, practices and plans for cities in the Global South would be flawed if contextual drivers of urban land use development and barriers to CCA and CCM are underestimated. Knowledge and good planning for addressing CC ought to address the problems of CC threats whilst bringing direct socioeconomic and environmental benefits to the poorer social groups whose livelihoods may be threatened by CCP, and who depend on the limited resources available, especially land rights to survive.

Unregulated informal land market and changes of use with the continued proliferation of informal settlements further compound the problem. This includes the undermining of food production and the decreasing income of many poor households. From the case of the AMP, it was clear that the proposals sharply contrasted the subsistence land use and development dynamics; firstly, the AMP appears to have been largely conceived and executed technocratically with little recognition of local specificities such as socio-cultural, economic and institutional contexts. The planning ideas bring to the fore what has been repeatedly observed: the over-emphasis of technical niceties of plans and the disregard of problematic socio-cultural and economic forces that, in fact, drive urban land development and undermine CCA and CCM initiatives. The tendency to see CCP as a non-local issue but as a global agenda instead is a real impediment to meaningful planning and budgeting management at the local level. The problem is further compounded by limited capacity/knowledge and will among bureaucrats, LGAs, officials, and central government policymakers.

The establishment of institutional mechanisms for ensuring CCP and structures for auditing and reporting is equally necessary. The current tendency towards centralised governance in decision-making must be re-examined to give voice to those at the levels where CC hazards are most felt and impacts most apparent. City/municipal level officials including bureaucrats/planners must be held accountable for what their specific authorities/units have done and achieved to plan for, prepare, and implement local adaptation/mitigation plans. They also must collaborate with other stakeholders to implement land use plans and take other measures to reduce greenhouse gas emissions and reduce the vulnerability of urban settlements to CC hazards. This includes the adaption of community/local experiences and practices that have directly or indirectly contributed to CCA and mitigation. Unregulated informality is the key driver of the current land use changes and abuses in the city. But the AMP land use proposals, for instance, in the peri-urban areas have underestimated informality, diverse actors, and interests. The AMP proposal seems to oversimplify the complexities of real urbanisation in the Global South; the master planners simply pushed informality away in lieu of middle-income housing. One of the measures to take according to the literature and local realities is to treat informality first as a solution, and then strive to nurture and adapt it to CC imperatives (Kombe 2017).

The need for institutional mechanisms to promote horizontal and vertical coordination among stakeholders and institutions involved in various matters that concern land use and CCA and CCM conservation at various levels cannot be overemphasised. The capacity to adapt to CC cannot be achieved without coordination at the city and community level and among sitting land occupiers. Affirmative and coordinated actions to work with the grassroots communities, integrated planning of the larger part of the city coupled with empowerment of local governments are necessary measures to institutionalise CC knowledge adaptation into spatial planning and implementation.

References

Bedsworth L, Hanak E (2013) Climate policy at the local level: insights from California. Global Environ Change 23(3):664–677. https://doi.org/10.1016/j.gloenvcha.2013.02.004

Biesbroek GR, Swart RJ, van der Knaap WG (2009) The mitigation–adaptation dichotomy and the role of spatial planning. Habitat Int 33:230–237

Burby RJ (1998) Cooperating with nature: confronting natural hazards with land-use planning for sustainable communities. Joseph Henry Press, Washington, DC

Carmin J, Roberts D, Anguelovski I (2011) Planning climate resilient cities: early lessons from early adapters. In: Fifth urban research symposium on cities and climate change: responding to an urgent agenda. World Bank, Marseille, France

Dodman D, Kibona E, Kiluma L (2011) Tomorrow is too late: responding to social and climate vulnerability in Dar es Salaam, Tanzania. Case study prepared for Cities and Climate Change. Global Report on Human Settlements

Feiock R, Kassekert A, Berry F, Yi H (2009) Institutional incentives and early adoption of sustainable energy innovations. In: Annual meeting of the American political science association. Toronto. https://papers.ssrn.com/sol3/papers.cfm?abstract_id=1450809&download=yes

Herslund L, Lund HD, Jørgensen G, Mguni P, Kombe WJ, Yeshitela K (2015) Towards climate change resilient cities in Africa—initiating adaptation in Dar es Salaam and Addis Ababa. In: Pauleit S, Coly A, Fohlmeister S, Gasparini P, Jørgensen G, Kabisch S, Kombe W, Lindley S,

Simonis I, Yeshitela K (eds) Urban vulnerability and climate change in Africa: a multidisciplinary approach. Springer International Publishing, Switzerland, pp 197–228
Hirokawa KH, Rosenbloom JD (2013) Land use planning in a climate change context research. In: Verschuuren J (ed) Handbook on climate adaptation law. Drake University Law School Research Paper, pp 12–33
John R, Jean-Baptiste N, Kabisch S (2014) Vulnerability assessment of urban populations in Africa: the case of Dar es Salaam-Tanzania. In: Edgerton E, Romice O, Thwaites K (eds) Bridging the boundaries: human experience in the natural and built environment and implications for research, policy, and practice. Advances in People-Environment Studies, Hogrefe, Cambridge, MA, pp 233–244
Kelman I, Gaillard J, Mercer J (2015) Climate change's role in disaster risk reduction's future: beyond vulnerability and resilience. Int J Disaster Risk Sci 2015(6):21–27. https://doi.org/10.1007/s13753-015-0038-5
Kithiia J (2010) Old notion–new relevance: Setting the stage for the use of social capital resource in adapting East African coastal cities to climate change. International Journal of Urban Sustainable Development 1(1–2):17–32
Kiunsi R (2013) The constraints on climate change adaptation in a city with a large development deficit: the case of Dar es Salaam. Environment and Urbanisation 25(2):321–337
Kombe W (2017) Institutional reforms for urban land use planning and management in Sub-Sahara Africa: evidences, gaps and prospects. E&D Vision Publishing Ltd., Dar es Salaam
Krause R (2010) Policy innovation, intergovernmental relations, and the adoption of climate protection initiatives by U. S. Cities. J Urban Aff 33(1):45–60. https://doi.org/10.1111/j.1467-9906.2010.00510.x
Krause R (2013) The motivations behind municipal climate engagement: an empirical assessment of how local objectives shape the production of a public good. Cityscape: J Policy Dev Res 15(1):125–141
Kweka T (2017) Land use changes in planned, flood prone areas the case of Dar Es Salaam City. Institute of Human Settlement Studies. Ardhi University, Dar Es Salaam
Kwon M, Jang H, Feiock R (2014) Climate protection and energy sustainability policy in California cities: what have we learned? J Urban Aff 36(5):905–924. https://doi.org/10.1111/juaf.12094
Leichenko R (2011) Climate change and urban resilience. Current Opinion on Environmental Sustainability 3(3):164–168
Malele BF (2017) Mismatch between human settlements planning and disaster risk reduction initiatives: the case of Dar Es Salaam City, Tanzania. Institute of Housing and Settlement Studies. Ardhi University, Dar es Salaam
Moser C, Satterthwaite D (2008) Towards pro-poor adaptation to climate change in the urban centres of low- and middle- income countries. Global Urban Research Centre Working Paper No. 1
Moser S, Ekstrom J (2010) A framework to diagnose barriers to climate change adaptation. PNAS 107(51):22026–22031. https://doi.org/10.1073/pnas.1007887107
Ong A (2011) Hyperbuilding: spectacle, speculation, and the hyperspace of sovereignty. In: Roy A, Ong A (eds) Worlding cities: Asian experiments and the art of being global, 205–226. https://doi.org/10.1002/9781444346800
Opp S, Osgood J, Rugeley C (2014) Explaining the adoption and implementation of local environmental policies in the United States. J Urban Aff 36(5):854–875. https://doi.org/10.1111/juaf.12072
Pan African, START, Tanzania Meteorological Agency and Ardhi University (2011) *Urban poverty & climate change in Dar es Salaam*. Dar es Salaam, Tanzania.
Parnell S (2015) Fostering transformative climate adaptation and mitigation in the African City: opportunities and constraints of urban planning. In: Pauleit S, Coly A, Fohlmeister S, Gasparini P, Jørgensen G, Kabisch S, Kombe W, Lindley S, Simonis I, Yeshitela K (eds) Urban vulnerability and climate change in Africa: a multidisciplinary approach. Springer International Publishing, Switzerland, pp 349–367

Pitt D (2010) The impact of internal and external characteristics on the adoption of climate mitigation policies by US municipalities. Eviron Plann C Gov Policy 28(5):851–871
Roberts B (2010) Prioritizing climate change adaptation and local level resilience in Durban, South Africa. Environ Urban 22(2):397–413. https://doi.org/10.1177/0956247810379948
Rugai D, Kassenga GR (2014) Climate change impacts and institutional response capacity in Dar es Salaam, Tanzania. In Macchi S, Tiepolo M (eds) Climate change vulnerability in Southern African Cities. Springer Climate, 39–55. https://doi.org/10.1007/978-3-319-00672-7
Sakijege T, Lupala J, Sheuya S (2012) Flooding, flood risks and coping strategies in urban informal residential areas: the case of Keko Machungwa, Dar Es Salaam, Tanzania. Jamba: J Disaster Risk Stud 4(1). https://doi.org/10.4102/jamba.v4i2.46
Sharp E, Daley D, Lynch M (2011) Understanding local adoption and implementation of climate change mitigation policy. Urban Affairs Review 47(3):433–457. https://doi.org/10.1177/1078087410392348
Shemdoe R, Kassenga G, Mbuligwe S (2015) Implementing climate change adaptation and mitigation interventions at the local government levels in Tanzania: where do we start? Curr Opin Environ Sustain 13:32–41
Shemsanga C, Omambia A, Gu Y (2010) The cost of climate change in Tanzania: impacts and adaptations. J Am Sci 6(3):182–196
Shi L, Chu E, Debats J (2015) Explaining progress in climate adaptation planning across 156 U.S. municipalities. J Am Plann Assoc 81(3):191–202. https://doi.org/10.1080/01944363.2015.1074526
Smucker TA, Wisner B, Mascarenhas A, Munishi P, Wangui EE, Sinha G, Weiner D, Bwenge C, Lovell E (2015) Differentiated livelihoods, local institutions, and the adaptation imperative: assessing climate change adaptation policy in Tanzania. Geoforum 59:39–50
Stevens MR, Senbel M (2017) Are municipal land use plans keeping pace with global climate change? Land Use Policy 68:1–14. https://doi.org/10.1016/j.landusepol.2017.07.026
Storch H, Downes N, Katzschner L, Thinh NX (2011) Building resilience to climate change through adaptive land use planning in Ho Chi Minh City, Vietnam. In: Otto-Zimmermann K (ed) Resilient cities: cities and adaptation to climate: cities and adaptation to climate change proceedings of the global forum 2010, local sustainability 1, Springer Science+Business Media B.V, pp 349–363. https://doi.org/10.1007/978-94-007-0785-6_36
URT (United Republic of Tanzania) (2007) National disaster management policy. Dar es Salaam: Government Printers.
Vasi I (2006) Organizational environments, framing processes, and the diffusion of the program to address global climate change among local governments in the United States. Sociol Forum 21:439–466
Vedeld T, Kombe WJ, Kweka C, Hellevik S (2015) Multi-level governance, resilience to flood risks and coproduction in Urban Africa. In: Pauleit S, Coly A, Fohlmeister S, Gasparini P, Jørgensen G, Kabisch S, Kombe W, Lindley S, Simonis I, Yeshitela K (eds) Urban vulnerability and climate change in Africa: a multidisciplinary approach. Springer International Publishing, Switzerland, pp 287–318

Wilbard Jackson Kombe is a Professor at the Institute of Human Settlement Studies at Ardhi University in Tanzania. Kombe received his PhD from the Technical University of Dortmund. Since his graduation, he has extensively collaborated with various institutions, both locally and internationally, in research and training activities. His recent research work has focused on land management and urban governance, informal urbanisation in poverty, urbanisation and public policy analysis. He has been extensively involved in applied and basic research activities in numerous countries in Sub-Saharan Africa. This includes research activities undertaken in Tanzania, South Africa, Zimbabwe, Kenya, Uganda, Mozambique, Zambia, Namibia, Botswana and Ethiopia. He has also led a number of multi-disciplinary research teams and projects undertaken in Tanzania and other countries in the region.

Samwel S. Alananga Ph.D., is an economist whose research focuses on asymmetric property markets and applied geospatial science in sectors such as land, human settlements, and urban economics. Since 2008, he has been a lecturer at the School of Earth Sciences, Real Estate, Business and Informatics at Ardhi University. He also has extensive teaching experience in economics, property finance, government finance, and geographic and land information systems. Dr. Alananga has contributed over 20 research articles to international journals. Currently, he is the Risk Coordinator and Head of the Department of Business Studies at Ardhi University.

Chapter 8
Contributions of Local Authorities to Community Adaptive Capacity to Impacts of Climate Change; A Case Study of Sea Level Rise in Pangani Division, Pangani District

Dawah Magembe-Mushi and Ramadhani Matingas

Abstract Communities around the world are facing tremendous climate change impacts, among others, sea level rise. Local authorities try to respond to the impacts of climate change by implementing localised adaptive measures to improve capacity. Impacts of climate change are worse in developing countries especially the Sub-Saharans due to inabilities to implement adaptive measures with inadequate resources; hence hindering the ability of communities to adapt to the impacts of climate change. It is the purpose of this chapter to explore the contributions of a local authority as one of the corresponding authorities responsible for enhancing a community's adaptive capacity and its behaviour to climate change impacts. Interviews, mapping, observation and photographing, were conducted in the Pangani division in Tanzania. SPSS and QGIS software were used for analysis. It was found that sea level rise caused communities to suffer beach erosion, land inundation, saltwater intrusion, changes in fish availability, destruction of infrastructure and vegetation. The local authority implemented localised adaptive measures which improved adaptive capacity. These measures included construction and maintenance of seawall, planting and protecting mangroves and preparation of land-use plans. Also, an institutional arrangement involved different departments, committees and the community at large, with actors collaborating through the local authority in developing community adaptive capacity. This chapter recommends that the local authority should; enhance coordination among actors, make use of spatial analysis tools, mobilise resources and enhance community participation.

Keywords Climate change · Adaptive measures · Local authority · Pangani

D. Magembe-Mushi (✉) · R. Matingas
Ardhi University, Dar es Salaam, Dar es Salaam, Tanzania
e-mail: dimushi2000@yahoo.com

R. Matingas
e-mail: ramamatinga@gmail.com

G. Alem Gebregiorgis et al. (eds.), *Planning Cities in Africa*, The Urban Book Series,
https://doi.org/10.1007/978-3-031-06550-7_8

8.1 Introduction

Schrag et al. (2007) explain, the vast increase in human activities, such as deforestation and fossil fuel combustion, causes concentrations of greenhouse gases (GHGs) in the atmosphere. These gases trap radiation from the earth's surface, warming the earth, then causing global warming. The average global surface temperatures have risen about 0.7 °C since early in the twentieth century. The trapping of gas emissions leads to greenhouse effects. Human activities contribute to emissions causing climate change. These emissions spread throughout the globe affecting people everywhere in different magnitudes, leading to disruption of economies and affecting lives, at the cost of people, communities, and countries today, and even more in the future. Climate change impacts are now affecting every country on earth. People are facing tremendous impacts, such as changing weather patterns, extreme weather events and rising sea level (United Nations 2017).

Among the mentioned effects, the rise of the sea level is one of the most troubling aspects of climate change and is likely to persist in the future from the progressive melting of the glaciers in the arctic (Meier and Wahr 2002). In the last two centuries, the long-term global sea level change was recorded to be only a few centimetres per century. Since then, the rate of rising increased substantially, for instance, in the last 20 years, both satellite and coastal sea level data show the rate of sea level rise has increased to about three centimetres per decade (Union of Concerned Scientists 2013). Given this, sea levels are very likely to rise much more quickly during the twenty-first century than in the twentieth century and will continue to rise. This is mainly caused by two factors; expansion of seawater and the melting of ice from the arctic due to warmth (Australian Academy of Science 2015).

At the same time, the coastal zones have been seen to change profoundly since the last century owing to rapid urbanisation. Today, low-elevation coastal zones below 10 m contain about 10 per cent of the world population (McGranahan et al. 2007). Sea level rise is a serious threat to countries with high concentrations of population and economic activity in coastal regions. The biophysical effects of sea level rise on coastal regions are inundation, flooding, storm damage, wetland loss, erosion, saltwater intrusion, coral bleaching (sea water temperatures), ocean productivity changes, species migrations, and destruction of infrastructure (Dasgupta and Meisner 2009). Singapore for example, lies only 15 m above the mean sea level, with about 30 per cent of the islands less than 5 m above the mean sea level. Singapore's National Climate Change Secretariat states that global warming could contribute to Singapore's sea level increase by 9.5 m, leaving 745,000 Singaporeans' homes submerged (Asuncion and Lee 2017). In the USA, there are cities and towns which are found within an elevation that could be flooded by the rise of the sea level. These areas are home to people, tourist destinations, fishing grounds, natural landscapes, military bases, financial centres, beaches and boardwalks, but they are all at the risk of sea level rise (Union of Concerned Scientists 2013). Flooding events and erosion have already been experienced in communities along the California coast and there is compelling evidence that risks will increase in the future (Heberger et al. 2009).

Furthermore, in Southern Africa, sea level rise and human development are together contributing to losses of coastal wetlands and mangroves and are increasing damage from coastal flooding in many areas. Projections indicate increasing risks of coastal erosion, due to climate change and sea level rise. The effect will be exacerbated by the increase of human-induced pressures on coastal areas (IPCC 2014).

In Tanzania, as in other countries with coastlines, impacts of sea level rise are increasingly manifested in accelerating coastal erosion and saltwater intrusion in many parts of the coastal areas, and in some cases destroying mangroves due to strong sea waves. These impacts have been vividly observed in the coastal areas of Bagamoyo, Pangani, Rufiji, and Zanzibar. Other impacts include the submergence of small islands like Maziwi in Pangani and Fungu la Nyani in Rufiji; the destruction of coastal infrastructures (some beach hotels in northern Dar es Salaam, the Pangani seawall in Tanga), and some settlements along the coast. Thus, the sea level rise is one of the most challenging climate change issues which not only causes destruction of key coastal infrastructure and coastal livelihoods but also affects the local communities along the coasts, as well as the economic development in general.

8.2 Local Authorities' Climate Change Adaption Strategies

Local government authorities have been impacted somehow both directly and indirectly by the effects of climate change. The impacts have been differently felt specifically by the communities and the local government of a particular area (Commonwealth of Australia 2007). Also, the impacts of climate change manifest at the local level, but the responsibility for an adaptive action does not lie solely with the local government alone. Effective adaption requires coordinated national leadership to support decisions made at the local level (NCCARF 2012). The local levels have been more important in mainstreaming climate change adaption for three reasons. Firstly, the impacts are manifested locally, affecting local livelihood activities, economic enterprises, health risks, etc. Secondly, vulnerability and adaptive capacity are determined by local conditions. Thirdly, adaption activities are often best practised and observed at the local level (OECD 2009). Due to these three reasons, the local authorities become the foremost organ in managing the impacts of climate change. They are mandated to prepare climate change adoption plans from city to settlement levels through policy options, such as early warning systems, health system's preparedness and responses and urban/settlement planning/ land-use planning; housing improvements; asset management and protection (NCCARF 2012). The authorities deal with the protection of forests through tree planting in nearby hills for prevention of landslides and floods and the protection of mangroves along the coastal areas to enable them to act as a natural seawall that defends the land from storm surges. Climate change adaption should also include making existing and new infrastructure resistant to the effects of climate change. This is also the responsibility of the local authority (UN-HABITAT 2004).

In Tanzania, local authorities together with other actors are responsible for human settlement development and land-use planning that will improve and protect settlements of communities in climate change risk-prone areas. Other duties include enhancing land-use planning, promoting building standards to accommodate impacts of climate change, relocating of settlements from high-risk areas, and promoting and supporting effective land-use planning at all levels. According to the National Climate Change Strategy of Tanzania (2012), the President's Office—Regional Administration and Local Government (PO-RALG) is working closely together with the Local Government Authorities (LGAs) through their various departments, in collaboration with lined sectoral ministries, to implement the strategic interventions at the local level (URT 2012) That being the case, local authorities along the affected coastal areas have been taking various initiatives towards adapting to climate change impacted sea level rise and its related effects. Therefore, this chapter intends to explore the effectiveness of local authorities in facilitating communities' adaptive capacity to climate change impacts, specifically the sea level rise. It identifies and maps communities that are affected, explains measures taken by local authorities in developing community adaptive capacity to the impacts, and suggests adaptive measures to improve the capacity of local authorities in dealing with climate change impacts.

8.3 Strategies for Sea Level Rise Globally

As a global challenge climate change is one of the 17 Sustainable Development Goals (SDGs) of 2015–2030. In its goal number 13, the SDGs argue for urgent actions to combat climate change and its impacts. In its targets 1–3, the goal insists on strengthening resilience and adaptive capacity to climate-related hazards and natural disasters in all countries, integrating climate change measures into national policies, strategies and planning, and improving education, awareness, and human institutional capacity to mitigations, adaptation, and impact reduction to climate changes. As explained in the previous section, local and regional governments, especially in cities, are often on the frontline in dealing with the effects of climate change. It is vital that these authorities' capacities to deal with climate-related hazards and natural disasters are strengthened so that we can directly protect our communities, particularly the most vulnerable ones. Again, at the local level, leadership in the past has adopted bottom-up approaches in combating climate change and raising awareness, which has been seen to lead to impact reduction through early warnings. It is, therefore, essential that local governments, particularly in the most vulnerable cities, integrate climate change adaption and mitigation into urban and regional planning to reduce the emissions of our cities and increase their resilience to environmental shocks.

8.3.1 Structural and Non-Structural Strategies

Different countries have come up with different strategies in combating the impacts of sea level rise resulting from climate change. The state of California in the USA, has more than 2,000 miles of open coast and enclosed bays. The city of San Francisco has been more vulnerable to the impacts of climate change, experiencing natural hazards such as storms, sea level rise, and extremely high tide. Two major measures have been adopted in San Francisco and California in general to develop adaptive capacity (Herberger et al. 2009). These measures are structural and non-structural. The structural measures involve the construction of dykes and dunes, seawalls, bulkheads, groins, breakwaters, beach nourishment, and raising existing structures and newly constructed ones (roads and railways). The non-structural measure is abandoning property and land to move to less threatened areas. In 2014, the city implemented "Guidance for Incorporating Sea level Rise into Capital Planning in San Francisco: Assessing Vulnerability, Risk, and Adaption" (ArcadisUS et al. 2016). The guidance provides direction to all city departments on sea level rise and how to incorporate it into planning. Furthermore, Arcadis US et al., identifies the Bay Conservation and Development Commission (BCDC) which leads the Adapting to Rising Tides (ART) Program in partnership with the National Oceanic and Atmospheric Administration and the Metropolitan Transportation Commission. The ART Program is a collaborative planning effort to help the San Francisco Bay Area communities adapt to sea level rise and storm event flooding.

8.3.2 Local and Transnational Self-Regulatory Activities

In Asian cities, Dhaka serves as an example of a coastal mega-city with a population of over 12 million inhabitants prone to climate change impacts. About 50 per cent of the city's population live in low-lying areas where waterlogging and drainage congestion is common due to river floods and excessive rainfall. The city is engaging in mechanisms to solve these problems at the local and transnational level (Breitmeier et al. 2009). The people of Dhaka are predominantly relying on private, self-regulatory activities for flood management through transnational NGOs, private firms and CBOs that support the urban poor with food, water, medicines, temporary shelters and relief during floods (Breitmeier et al. 2009). The local government in Dhaka has engaged in a non-inclusive policy approach to various measures, and as such local communities are constrained from engaging with them. For example, the Greater Dhaka Flood Protection Project (GDFPP) was established in 1989 as a coordinated effort involving stakeholders in the city. A number of city structures were constructed or improved; these include embankments, flood walls, and infrastructure such as roads. Since the project was not inclusive, the embankments implemented saved people and properties from floods but also caused waterlogging and internal drainage congestion which becomes severe during heavy rainfall (Alam et al. 2007 cited in Breitmeier et al. 2009).

8.3.3 Regional Adaptive Strategies

Africa is one of the continents that faces a great impact from climate change. African countries' local authorities develop community adaptive capacity as the continent experiences severe effects due to the high level of poverty among its people. Mozambique's capital city Maputo has a geographical location that makes it highly vulnerable to the impacts of climate change such as sea level rise. The capital is the most densely populated urban area with more than 70 per cent of its total population living in areas prone to floods, soil erosion, and other environmental hazards (Dodman 2012). The effects of sea level rise are becoming more and more familiar to the extent that they require expensive coastal management by the Maputo City Municipality. These effects include saltwater intrusion within the land negatively impacting agricultural activities, thus adding to the already existing urban poverty. There is a great reduction in the number of sand strips along the beaches due to continuous movement of the sea creating significant coastal erosion and impacting negatively on economic activities such as tourism (UN-Habitat 2011). Flooding incidences are becoming frequent in Maputo. The "Living with Floods" initiative aims to protect lives and livelihoods by building elevated schools and community halls at low cost in flood-prone areas (UN-Habitat 2004). The city council created two new departments for environmental inspection and management within the existing framework to strengthen the role of the municipal authorities in the mitigation and adaption measures (UN-Habitat 2011).

8.4 Conceptualising Community Adaptive Capacity

Other authors describe community resilience as an important concept for characterising and measuring abilities of populations to anticipate, absorb, accommodate or recover from the effects of a hazardous event in a timely and efficient manner (Kruse et al. 2017). Kruse and others explain that the role of community is considered vital for building disaster resilience, yet community resilience as a scientific concept has been vaguely defined and lacks the guiding characteristics necessary for analysing and enhancing resilience on the ground. The embrace framework of community resilience (Fig. 8.1), as discussed in their paper on disasters published in *Natural Hazards and Earth Systems Sciences* provides an empirical tool for understanding, explaining, and measuring community resilience to natural hazards. The conceptual framework on community resilience is manifested across three principal areas: resources and capacities, actions, and learning. In addition, community resilience is also influenced by extra-community forces that govern disaster risk such as laws,

Fig. 8.1 Conceptual framework (*Source* Adopted and modified from Kruse et al. 2017, licensed under CC-BY 4.0)

policies and responsibilities on the one hand, and on the other, the general societal context, natural and human-made disturbances, and system change over time.

The framework is graphically portrayed in this way so that its application can assist in guiding the assessment of community resilience in a systematic way and identify the key drivers and barriers of resilience that affect a particular hazard-exposed community (Kruse et al. 2017). Community resilience surpasses the social-ecological understanding of resilience as it incorporates social factors such as perceptions, beliefs, institutional environment, and governance settings that influence or shape the capacity of communities to build resilience (Ensor and Harvey 2015). Other conceptual and empirical studies to date have indicated that community is an important scale at which to build resilience that can enhance both individual/household, as well as the broader population. Therefore, this chapter intends to explore the contribution of local authorities in building community adaptive capacity to the impacts of climate change's hazards and disasters, such as that of sea level rise which leads to flooding.

8.5 Methodological Approach

This chapter is based on explorative research responding to "what" and "how" local government contributed to building capacity for community resilience. In answering the how and what questions, a case study approach was considered an appropriate strategy to be used in researching how the local government contributed to community adaptive capacity to climate change in the Pangani district, Tanga region, in Tanzania (Fig. 8.2). Data collection methods involved household surveys in the community, official interviews with representatives of local government bodies responsible for disaster management, observations and mapping of the affected areas. Adaptive measures in the case study area were carefully and completely observed in social units. This chapter uses both qualitative and quantitative data. Photographic registration and mapping were used to describe and provide evidence on spatial information. Computer aided software such as SPSS was used for statistical data analysis and QGIS for processing spatial data.

8.5.1 Location of the Pangani Division

The Pangani division, locally known as Pangani Town, is one of the four divisions in the Pangani district, the others being Mkwaja, Madanga and Mwera. The division is

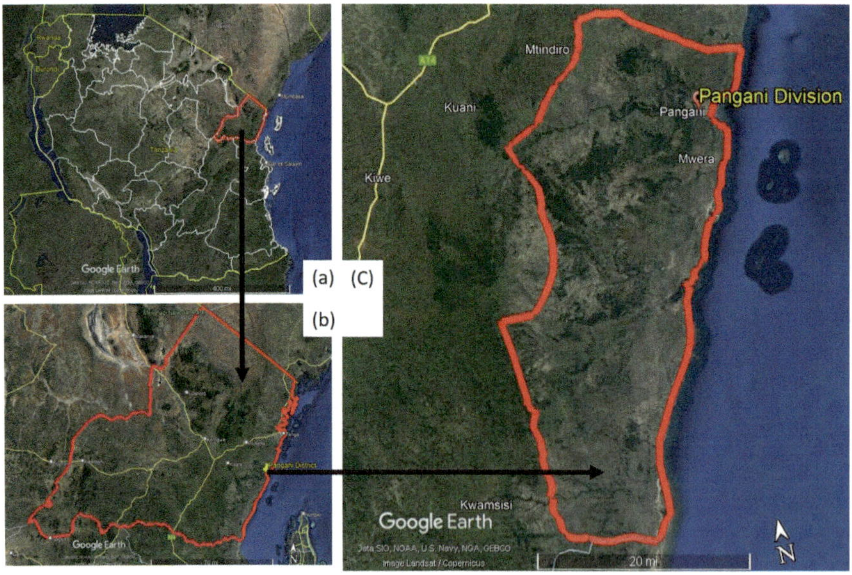

Fig. 8.2 Location of the Pagani division **a** Tanga region in Tanzania; **b** Pangani division in Tanga; **c** Pangani division (*Source* Authors' map on GoogleEarth 7.3, 2022)

located 45 km south of Tanga City, at the mouth of the Pangani River which is one of the largest rivers in Tanzania. The Pangani division is made up of three wards. These wards are Pangani Magharibi, Pangani Mashariki and Bweni. The division has 10 hamlets: Pangani Mashariki has the hamlets of Funguni, Mkoma, Malindi and Gombero, while Pangani Magharibi has Kumba, Sokoni and Kinara, and Bweni has Bweni, Kikokwe and Mashado. The division covers an area of 39.01 km^2 which makes up 2 per cent of the Pangani district and it is bordered by Mwera, Madaga and the Indian Ocean (Fig. 8.2).

8.5.1.1 Sampling Design

This chapter applies both probability and non-probability sampling. The probability sampling method was applied in obtaining household surveys, with a total of 77 households interviewed in the Pangani division. The sample size was derived through Slovin's formula as follows; $n=\frac{N}{(1+N\times e\times e)}$.
where: n = Sample population _ N = Total population _ e = Error tolerance.

Calculating for "n"

$n=\frac{10500}{(1+10500\times 0.05\times 0.05)}=385\ people$

The average household size of Pangani Division is 5 hence $\frac{385}{5}=77 Households$

Proportional sampling was again employed to obtain the number of households to be interviewed in each ward. For Pangani Magharibi, Pangani Mashariki and Bweni there were 46, 22 and 9 households respectively.

The non-probability sampling was applied through purposive sampling. The selected interviewees were principally government officials and local leaders which included: Natural Resource and Environmental Officers from the Pangani District Council, the Tanzania Meteorological Agency, and the Ward Executive Officers (WEO) from the three wards of the Pangani Division mentioned above.

8.6 Climate Change Impacts Affecting the Pangani Division

In Pangani, it was observed that the main climate changes that are affecting the division are: change in rainfall patterns, increase in temperature, and sea level rise. The change in rainfall patterns was through varying rainfall seasons which are no longer predictable and are decreasing from three main rainfall seasons to two or just one per annum. These changes also cause the rainfalls to become either shorter or longer than expected, but in most cases, they are shorter and of high intensity. Change in rainfall patterns affects the wellbeing of the people in the Pangani division socially and economically, as recent crops such as coconuts hardly survive in Pangani due to the shortage of rainfall. Increase in temperature across Pangani and Tanga was also observed to have relatively fewer fluctuations throughout the year and an increase in temperature has been witnessed in most parts of the country and especially in coastal areas. Being located on the coast Pangani division also faced a temperature increase which impacted the ecosystem in Pangani and its residents since it led to heat stress. The annual long-term average temperature over the last 20 years ranges from 30.9 °C (maximum) to 23.1 °C (minimum). According to Tanzania Meteorological Agency (2018), over the last 20 years Pangani division and the Tanga region at large has experienced an increase in temperature of about 1 °C. This increase has effects on the environment, particularly the ocean, as the temperature increase in the surrounding area triggers seawater to expand more and more, which in turn causes sea level rise.

8.6.1 Sea Level Rise (SLR)

Sea Level Rise (SLR) is one of the recent and major climate change impacts witnessed in most of the coastal communities including the Pangani division. SLR is a serious threat to the Pangani division, and its effects can easily be seen all along the coast. The effects of the sea level rise became more alarming in Pangani when Maziwi Island began to submerge and erosion became prominent on natural features, as was seen when trees were eroded and washed away into the sea. Also, it was observed that the sea wall which was constructed during the colonial era had eroded badly due to the force of the high tides that had never been experienced before. According to one elderly respondent:

> Many years ago, people used to cross over to Bweni by foot but as days go by the means of crossing over changed mainly due to technology improvement and the rise of sea level. We were using floating rafts but since land inundation happened and erosion of riverbanks, you can't cross over by foot anymore.

According to Elasha-Osman (2006), the projected subsidence rates due to sea level rise are 15–95 cm from 2006 to the year 2100, where the rate of rising increases from 0.2 cm/annum to 1.1 cm/annum. Adapting the average rate of sea level rise, which is 0.65 cm/annum, a sea level rise of 53.3 cm by 2100 will inundate approximately 2 km^2 of the land in the Pangani division if no measures are taken in the adaptability of the division and district in general, even though this is approximately 5.3 per cent of the division land coverage. If no measures are devoted to addressing this, with this inundation infrastructure such as roads will be destroyed, coastal vegetation will be washed away, most beaches will vanish, and dozens of people will be left homeless.

8.6.2 Communities Affected by Sea Level Rise in Pangani Division

It was observed that Pangani division is the most affected division among the four divisions in the Pangani district, mainly due to its geographical location and topographical characteristics. Approximately 24 per cent of Pangani division is located below sea level (altitude of about −9 m). Literature indicates that the most troubling effect of SLR is flooding. However, there are also other effects of SLR such as beach erosion and destruction of infrastructure and vegetation, saltwater intrusion, land inundation and waterlogging. The communities of all of the ten hamlets in the division were affected differently depending on their proximity to the sea, the human activities conducted, and the population density. For example, the Mashado hamlet in Bweni ward does not directly experience the effects of SLR such as beach erosion, because it is not located along the Indian Ocean, hence the effects that were observed to prevail in Mashado and other hamlets away from the ocean included saltwater intrusion and flooding.

8.6.2.1 Beach Erosion

The Pangani division has a shoreline of about 13 km and approximately 80 per cent of it was observed to be affected by erosion. Erosion occurs mainly during high water tides when seawater hits the surface and washes away sand from the beaches. The loss of sand causes the beach to become narrower and lower in elevation. Beach erosion is evident in areas such as Kumba in Pangani Magharibi, Pangadeco in Pangani Mashariki and Kikokwe in Bweni (Fig. 8.3). Pangani division relies heavily on beach-related tourism as one of the major sources of revenue for the division and district in general. Now touristic activities are threatened due to beach erosion. Beach erosion also threatens coastal properties, buildings housing homes and businesses, and infrastructure such as roads.

8.6.2.2 Flooding/Waterlogging

Literature shows that areas that are facing SLR are prone to flooding/waterlogging during rain seasons mainly because SLR is accompanied by a rise in the water table of an area, and when the water table rises, the retention capacity of soil becomes high. The situation worsens if the area has poor or ineffective stormwater drainage, which leads to the likelihood of flooding. The Pangani division is currently experiencing the problem of waterlogging in streets and other areas for a much longer time than usual. Although this is little known to most of the inhabitants, according to much of the literature this phenomenon is caused by SLR.

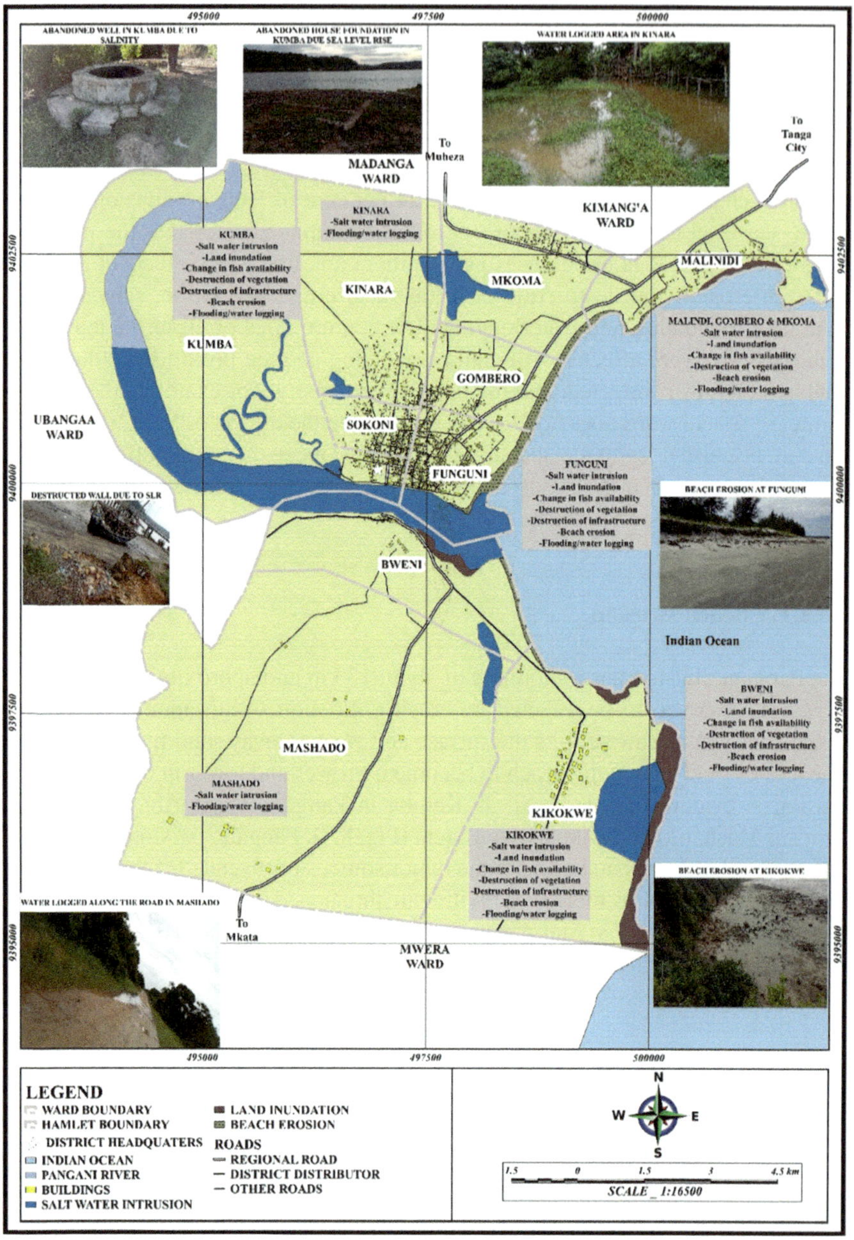

Fig. 8.3 Effects of sea level rise in the Pagani district (*Source* Fieldwork, May 2018)

8.6.2.3 Saltwater Intrusion

Saltwater intrusion affects approximately 69 per cent of the population in Pangani division. Salt water is seeping into the land, making it more saline which in turn makes the cultivation of crops difficult. Salt water is also getting into shallow wells/aquifers making their water saltier so that some of them are being abandoned (Fig. 8.3). The Pangani division also experiences crop failure for some crops such as coconuts, maize, cassava and banana in some areas caused by the increase in land salinity. This crop failure threatens the food security of the division and the district at large.

8.6.2.4 Destruction of Infrastructure

The infrastructures in Pangani division are at risk of destruction due to SLR. These infrastructures are roads, the seawall and buildings. The seawall that was constructed by the Germans in the late nineteenth century is to a great extent destroyed, primarily because seawater has come closer and the tides have become stronger, easily weakening the sea wall so that some parts began to collapse. Roads and buildings are destroyed mainly due to continuous inundation and erosion which is a result of SLR.

8.6.2.5 Destruction of Vegetation

Both inland vegetation such as conifer (locally known as mivinje), and marine vegetation such as mangrove (locally known as mikoko) are in great danger of destruction. This destruction will impact both biological and cultural resources located along the coast. In the Pangani division, approximately 8 ha of coastal vegetation has been lost due to prolonged and continuous erosion mainly caused by tidal water hitting the surface and eroding the soil (Fig. 8.3).

One fisherman from Kumba stated that:

> Some years ago, this place had trees all over and the fishing camp wasn't at this spot, but with time, trees were washed away, one by one, and now there is nothing, it's only water as if nothing ever existed here.

8.6.2.6 Land Inundation

As the sea level rises, more land is being inundated by the sea and this has implications for the availability of land to put to various purposes such as agriculture and settlement development. In the Pangani division land inundation is evident in Kumba, Bweni, Kikokwe and Funguni where to date approximately 1 km^2 of land has been inundated (Fig. 8.3). As SLR persists, land inundation will continue to occur resulting in the inundating of approximately 200 ha (2 km^2) of land by 2100 if no measures to tackle the issue are implemented before then. This will result in the destruction of infrastructures such as roads, coastal vegetation will be washed away, most beaches will have vanished, and many people will be left homeless.

8.7 Local Authority and Community Adaptive Capacity

It has been revealed that local authorities formulated various measures to develop and increase community adaptive capacity through both structural and non-structural measures. Among the various local authorities, the Pangani District Council (PDC) is taking measures to neutralise the impacts of SLR to its communities by developing adaptive capacity. These measures were categorised into two types: the structural measures and the non-structural measures. This section discusses how the PDC builds adaptive capabilities socially, economically and culturally according to the conceptual framework of this chapter.

8.7.1 Actions

8.7.1.1 Structural Measures

These types of measures are those that were observed as they were physically seen in the Pangani district. These measures include construction and maintenance of sea walls, planting and protecting mangroves and construction and maintenance of stormwater drainage. For some time, the PDC has been doing periodic maintenance of the existing seawall, which is wearing out fast, having existed since the 1890s (Fig. 8.4). This situation raised fear among the Pangani residents of being engulfed by the sea until the PDC asked for support from the Central Government via the concerned ministry.

The sea wall was under construction, and it involved two sides; the Pangani Mashariki and the Pangani Magharibi with 1,145 m (900 m completed), and the Bweni side with 665 m, on which the construction was yet to start (Fig. 8.5). Milman (2018) makes the case for planting mangrove forests as they are coastal vegetation considered crucial buffers to storms and saltwater intrusion and provide habitats for marine creatures. Therefore, mangroves are crucial for curbing the effects of SLR since they provide natural protection to coastal areas because they grow in salty water, maintain land stability and reduce the force of tides. In the Pangani division it was observed that mangroves were planted, and they were highly protected by the communities themselves through the Beach Management Unit (BMU) (Fig. 8.5). Along the Pangani River just beside the seawall, 21,000 mangrove seedlings have been planted across 10 ha of land, and more are expected to be planted in the division and the district in general (Fig. 8.4). However, in recent years, due to human activities mangrove forests have been cut down, but the PDC has been working closely with the community through the BMU to restore them (Fig. 8.4). With respect to stormwater

a) New seawall along Pagani River in Kumba b) Old seawall along Pagani River in Bweni

c) Planted mangrove trees along the seawall d) Stormwater drainage along Jamhuri road

Fig. 8.4 Structural measures in the Pagani division (*Source* Fieldwork, May 2018)

drainage, the PDC through its department of works, worked closely with the Tanzania Rural and Urban Roads Agency (TARURA) to ensure that the streets were well drained by the construction of storm water drainage channels and their associated structures such as culverts (Fig. 8.5).

8.7.1.2 Non-Structural Measures

Adaptive capacity was also increased by the PDC in collaboration with other stakeholders. Through the action of awareness creation, education was provided to the Pangani residents on climate change impacts and their associated effects, as well as various environmental conservation issues. Awareness was created by conducting seminars and focus group discussions with various groups, such as fishermen, women and youth. It was found that the PDC has provided training to 65 technical officers on climate change impacts and how best to develop community adaptive capacity in a participatory way. Also, the PDC has provided training to all the BMU members of *Pangani Mashariki*, *Pangani Magharibi* and *Bweni* on climate change impacts and how to restore mangrove vegetation. Radio broadcasting was used in creating awareness; through it, education on environmental issues, illegal fishing and other concerns was provided, and citizens had an opportunity to call and ask questions of the experts involved. These are the actions that were taken by the PDC as a responsible local

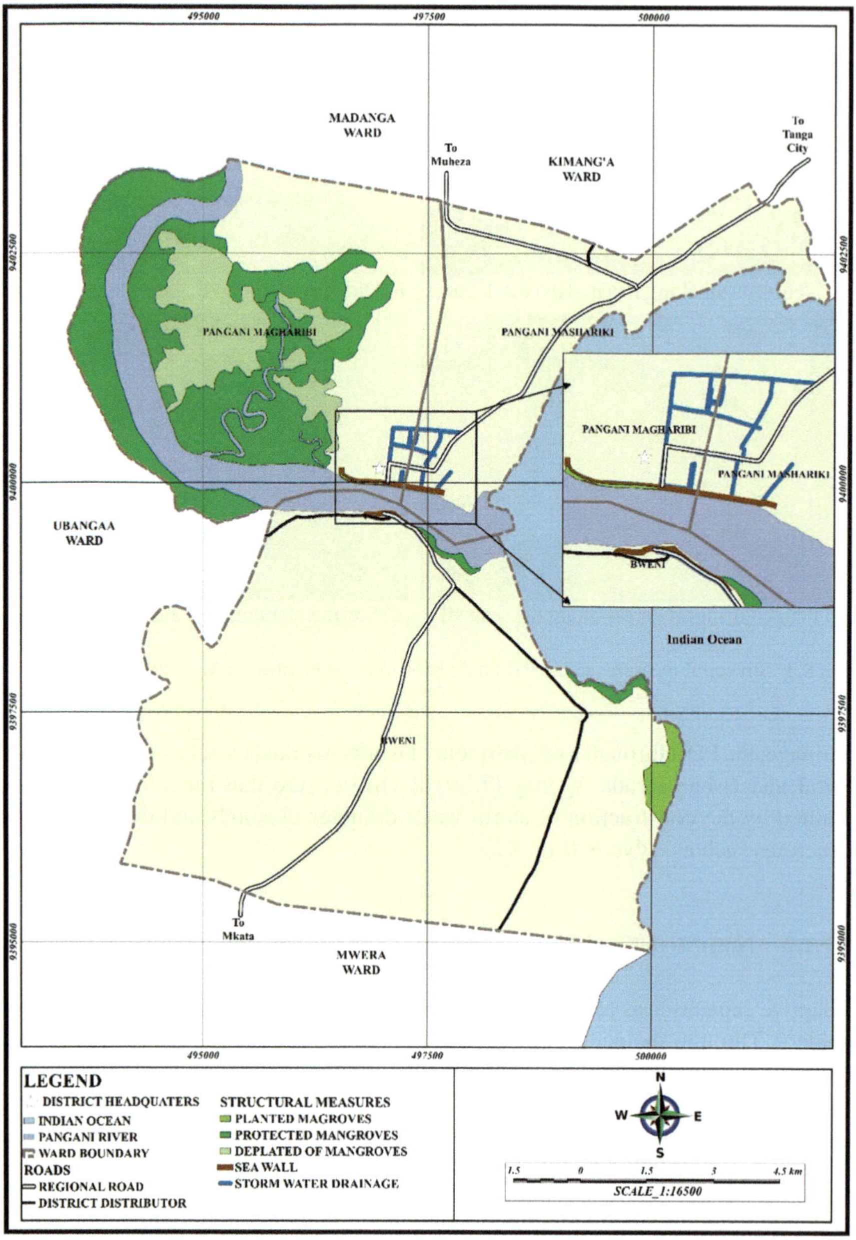

Fig. 8.5 Actions for community adaptive capabilities in the Pagani division (*Source* Fieldwork, May 2018)

authority so as to build adaptive capacity. The following section discusses resources in relation to capacity which is the second component in the conceptual framework that the PDC has in building the adaptive capacity for the Pangani community.

8.7.2 *Resources and Capacity*

Through literature and evidence from empirical data, it was found that resources are important components in developing community adaptive capacity to climate change impacts. These include human, technical and financial resources, which were applied in implementing both structural and non-structural measures. The PDC had the following resources and capacity in its process of developing community adaptive capacity to sea level rise as an impact of climate change.

8.7.2.1 Human Resources

This can be termed as the manpower, either skilled or unskilled, which can be used in executing different actions aimed at developing community adaptive capacity. Skilled manpower was such intellectuals as PDC staff, local leaders such as WEOs and other stakeholders from different organisations such as the Climate Action Network Tanzania (CAN TZ) and the Pangani Heartland Development Organization. These human resources and actors participated in different activities such as education provision and preparation of land-use plans, which in turn resulted in developing community adaptive capacity to climate change impacts within the division. Community members acted as both skilled and unskilled manpower. For example, manpower to implement various programmes, such as the restoration of mangrove forests, was obtained by mobilisation and sensitisation of community members through the BMUs. Moreover, this type of resource was applied in education provision, and consultancy services concerning climate change impacts and how best to adapt. Human resources of the local authority faced challenges since those available were not sufficient to address all the required activities that were scheduled to be performed by the PDC.

8.7.2.2 Financial Resources

Financial resources are also known as monetary resources that are used to implement actions both structural and non-structural to develop community adaptive capacity. The main sources of these resources are donor agencies, in-kind contributions, central government, NGOs and the PDC funds through the collection of taxes and levies. For example, resources that were used for the construction of the seawall were obtained through the United Nations Environmental Programme (UNEP) sponsorship with a total cost of approximately TZS 2.4 billion under the programme of Addressing

Climate Change Impacts on key infrastructure and settlements. However, as a local authority, the PDC has insufficient funds due to its low capacity in collecting taxes and levies within the district.

8.7.2.3 Technical Resources

Technical resources are equipment and expertise used or facilitating the implementation of adaptive measures. These resources include boots, hand hoes, laptops, projectors, radio calls, mobile phones, vehicles, etc. These resources facilitated actions such as patrol vehicle use as a technical resource and planting mangroves using boots and hand hoes. The sources of technical resources came from the PDC, central government, and actors such as CAN TZ, Saadani National Park (SANAPA), Sea Sense, and others who were interested in a particular programme or project. Again, according to official interviews with the PDC environmental officer, though resources were declared as scarce since impacts of climate change are addressed in a multidisciplinary way, the local authority was implementing their plans by collaborating with different stakeholders and sponsors.

8.7.3 Learning

Learning is the third and last component of community resilience to climate change identified in the conceptual framework. In this component variables that were considered are risk or loss perceptions, how the risk or loss is problematised by the community, the reflections of the risk/loss according to the local community, experimentation and innovation with adaptive measures, dissemination, monitoring and review of the strategies that can be, or that were used, in dealing with climate change effects and becoming resilient.

8.7.3.1 Experimentation and Innovation

In the Pangani division, the main actors conducting experiments and innovation are various NGOs which operate locally and nationally. One of these NGOs is Sea Sense which operates nationally and aims at conserving and protecting marine wildlife such as turtles, dugongs, whales, dolphins and whale sharks. This marine wildlife is endangered mainly due to climate change and human activities. Sea Sense carries out different experiments and projects: fisheries management, education and outreach, sea turtle nest monitoring, cleaning up marine debris and many others. Furthermore, Sea Sense organises different community events each year to educate and create awareness on issues related to the impacts of climate change. These events usually take place during special events such as World Environment Day, World Oceans Day, World Sea Turtle Day and World Fisheries Day. These events combine entertaining games and challenges with more focus on conservation activities such as beach cleaning and tree planting. The other NGO that engages with experimentation and innovation is, CAN TZ which works closely with the PDC in creating and supporting small environmental projects which aim for climate change resilience within the Pangani district. Through these stakeholders the PDC was able to create experiments which led to innovation and different ideas and methods to deal with the sea level rise and other impacts of climate change within the district.

8.7.3.2 Dissemination

Dissemination of information in the Pangani division is through different institutions, organisations and individuals. The first form of dissemination is from the government institutions/organs which provide directives and information. This mainly involves the Vice President's Office (VPO), the office of the Regional Administrative Secretary of Tanga (RAS-Tanga) and the PDC. The VPO through the ministry involved works closely with the PDC in ensuring climate change impacts are dealt with effectively. The ministry is responsible for providing instructions and guidance on climate change adaption strategies to the PDC and for funding the implementation of projects and programmes. RAS-Tanga acts as a bridge between the PDC and the VPO. The second form of dissemination involves governmental institutions, NGOs, and the community at large. In this form, the PDC works closely with organisations such as CAN TZ, Sea Sense, FORUM-CC, Pangani Heartland Development Organization (PHDO), the University of Zanzibar, Saadani National Park, and others in providing information and creating awareness on climate change issues to communities. Additionally, an NGO known as Uzima Kwa Sanaa (UZIKWASA) operates Radio Pangani FM (107.7 MHz) providing sessions to the PDC every Saturday at 17:00 h creating awareness of environmental issues, i.e. illegal fishing and other related issues. Afterwards community members are able to call in and ask questions of the officials. The third form of dissemination involves the community members through their local leaders such as the WEO, VEO, Village Chairperson and the BMU members. These local leaders usually inform the community members of the DOs and DON'Ts that can help in dealing with the impacts of climate change. They also convey messages from the PDC.

8.7.3.3 Risk/Loss Perception

In the Pangani division, risk/loss is mainly attributed to the described impacts of sea level rise. The community in the Pangani division perceived the beach erosion that affected 80 per cent of the shoreline as the most disturbing effect of sea level rise, which troubled 89 per cent of community members. Furthermore, land inundation was perceived as the second most troubling effect of sea level rise in the Pangani division, from which 80 per cent of the population was affected, and feared that Pangani Town might be engulfed, especially once the old seawall was destroyed.

8.7.3.4 Monitoring and Review

There were different actions which were taken to ensure communities in the Pangani division adapt to the ongoing phenomenon of climate change. This sequence of actions needed to be monitored and reviewed over a specific timeframe. Monitoring and review in Pangani division are mainly done by the government through the VPO and the PDC to help in determining whether the planned outputs and outcomes have been achieved. For example, the VPO set the FORUM-CC and the University of Zanzibar the task of creating awareness on climate change issues in the communities of the Pangani division and the district at large. To ensure that the intended outcomes and outputs of the task were achieved, the VPO in collaboration with the PDC conducted monitoring and review by receiving and analysing progress reports of the consultancy service, field observation, focus group discussions with communities, and baseline measurements.

8.8 Conclusions and Recommendations

This section summarises the conclusions and recommendations of this study based on the objectives and variables of the research.

8.8.1 Conclusions

The intention of this chapter was to explore the contribution of the local authority in developing community adaptive capacity to impacts of climate change, specifically sea level rise, as one of the critical climate change impacts facing the Pangani division. A number of effects were found that communities faced because of sea level rise. These effects threatened the sustainability of the ecosystem in the Pangani division and in turn the communities' welfare was disturbed socially, economically and culturally. The PDC as the local authority involved came up with different approaches, strategies, and projects for developing community adaptive capacity to climate change impacts (sea level rise). According to the community resilience framework, the PDC undertook various adaptive actions, both structural and non-structural measures, aiming for preparedness, recovery, and mitigation. The local authority also provided learning mechanisms that were achieved through participation of the community by way of different activities and decision-making in the Pangani division. Findings show that about 75 per cent of the households participated in different adaptive measures in Pangani. This had positive results in developing community awareness; learning from the measures the adaptive capacity to climate change impacts. In terms of resources and capabilities of the community, the study found that despite the hurdle of inadequate resources that the local authority faced, they worked with different actors at different levels, as well as the community at large, to develop capacity for adaptive mechanisms. The local authority worked closely with the central government via the VPO and integrated other actors such as FORUM-CC, UZIKWASA, Sea Sense, UNEP, the University of Zanzibar, SANAPA and CAN TZ. They also provided human, financial, and technical resources and conducted different adaptive measures that contributed to community adaptive capacity to climate change impacts. These actors were challenged by poor coordination since each actor worked independently with the local authority and not with one another for collaborative efforts.

8.8.2 Recommendations

The following recommendations arose from the emerging issues, findings and analysis: it is recommended to the local authority concerned and others of the like, to coordinate different stakeholders available so as to create a better community adaptive capacity to climate change impacts.

8.8.2.1 Beach Nourishment

According to various literature, beach nourishment is among the measures used to curb beach erosion. In San Francisco, beaches were highly eroded and failed to perform their functions until the local authority in collaboration with other actors undertook beach nourishment. It was a successful measure and the beaches were restored. The same could be applied in the Pangani division to revive the eroded beaches of Kumba, Pangadeco and Bweni. The local authority could implement this measure by collaborating with actors such as VPO, SANAPA and UNEP.

8.8.2.2 Using Spatial Analysis Tools

In developed countries such as the UK, local authorities use spatial analysis tools for the provision of stormwater drainage in rural and urban areas and succeed in establishing sustainable stormwater drainage that contributes to community adaptive capacity to climate change impacts. This measure could also be undertaken in the process of planning and designing stormwater drainage systems as one of the structural measures for dealing with the impacts of sea level rise. This would facilitate the alignment of drains according to the terrain to enhance the functionality of the stormwater drainage system; thus, leading to community adaptive capacity. This measure can be implemented by the PDC, the department of works in collaboration with TARURA using GIS software, Global Mapper and others.

8.8.2.3 Resource Mobilisation

Management theory argues that it is important to have human and other resources to conduct activities of planning, organising, actuating and controlling performance to determine and accomplish stated objectives. Resources help in implementing projects and programmes which assist in developing community adaptive capacity to climate change impacts. The PDC can mobilise resources through maximising efficiency and effectiveness in collection of taxes and levies, searching for donors and sponsors, the reallocation of funds, developing other sources of funds through tourism, for example, integrating more actors, and community sensitisation.

8.8.2.4 Enhance Community Participation

Community participation at all levels should be enhanced in the process of developing adaptive capacity to climate change impacts as it greatly assists in curbing the challenge of the scarcity of resources. This can be done by the departments of environment and sanitation as well as through community development and youth by collaborating with Sea Sense, CAN TZ and SANAPA.

8.8.2.5 Coordination of Actors

Actors that are responsible for developing community adaptive capacity to climate change should be coordinated with one another rather than working independently with the local authority and other institutions. The proper coordination of key actors will lead to cumulative efforts for better development of community adaptive capacity to climate change impacts. This can be done by the PDC, and the District Executive Director by involving all the actors and discussing how best they can collaborate and coexist to develop community adaptive capacity.

References

ArcadisUS Inc., CallisonRTKL, Wageningen University and Research Centre (2016) Sea level rise adaption study: waterfront strategies for longterm urban resiliency. San Fransisco.

Asuncion RC, Lee M (2017) Impacts of sea level rise on economic growth in developing Asia. Asian Development Bank, Mandaluyong City

Australian Academy of Science (2015) The science of climate change: questions and answers. Australian Academy of Science, Canberra

Breitmeier H, Kuhn J, Schwindenhammer S (2009) Analyzing urban adaptation strategies to climate change: a comparison of the coastal cities of Dhaka, Lagos and Hamburg. Fern Universität, Hagen

Commonwealth of Australia (2007) Climate change adaption actions for local government. Australian Greenhouse Office, Canbera

Dasgupta S, Meisner C (2009) Climate change and sea level rise. The World Bank Environmental Department, Washington, DC

Dodman D (2012) Cities and climate change initiatives.UN-HABITAT, Nairobi
Elasha-Osman B (2006) Background paper for the African workshop on adaptation implementation of decision 1/CP.10 of the UNFCCC convention. UNFCCC, Accra, Ghana
Ensor J, Harvey B (2015) Social learning and climate change adaptation: evidence for international development practice. Wires Clim Change 6:509–522. https://doi.org/10.1002/wcc.348
Heberger M, Cooley H, Herrera P, Gleick P, Moore E (2009) The impact of sea level rise in the California Coast. Pacific Institute, Oakland
IPCC (International Panel on Climate Change) (2014) Climate change 2014 synthesis report summary for policymakers. Geneva
Kruse S, Abeling T, Deeming H, Fordham M, Forrester J, Jülich S, Karanci AN, Kuhlicke C, Pelling M, Pedoth L, Schneiderbauer S (2017) Conceptualizing community resilience to natural hazards—the emBRACE framework. Nat Hazards Earth Syst Sci 17:2321–2333. https://doi.org/10.5194/nhess-17-2321-2017
McGranahan G, Balk D, Anderson B (2007) The rising tide: assessing the risks of climate change and human settlements in low elevation coastal zones. Environ Urban 19(1):17–37. https://doi.org/10.1177/0956247807076960
Meier M, Wahr J (2002) Sea level is rising: do we know why? Proceedings of the National Academy of Sciences of the United States of America, Washington, DC
Milman O (2018) Everglades under threat as Florida's mangroves face death by rising sea level. The Guardian. https://www.theguardian.com/us-news/2018/may/02/mangroves-everglades-florida-rising-sea-level. Accessed 5 June 2018
NCCARF (National Climate Change Adaption Research Facility) (2012) Challenges of adaptation for local governments. Southport
OECD (Organization for Economic Co-operation and Development) (2009) Integrating climate change adaptation into development co-operation. Paris
Schrag D, Wofsy S, McCarthy J, Moorcroft P, Bloom D, Spengler J, Pringle A (2007) The habitable planet: a systems approach to environmental science. Annenberg Foundation, Los Angeles
Tanzania Meteorological Agency (2018) Data delivery report. Ministry of Works, Transport and Communication. Tanzania Meteorological Agency, Dar es Salaam
UN-Habitat (2004) Local leadership for climate change action. UN-HABITAT, Nairobi
UN-Habitat (2011) Cities in climate change initiaives. UN-HABITAT, Nairobi
Union of Concerned Scientists (2013) Causes of sea level rise. Cambridge
UN (United Nations) (2017) The sustainable development goals report. United Nations, New York
URT (United Republic of Tanzania) (2012) National climate change strategy. Minister of State Vice President's Office, Dar es Salaam

Dawah Lulu Magembe-Mushi , Ph.D., is Acting Director of the Centre for Continuing Education and senior lecturer at the School of Spatial Planning and Social Sciences at Ardhi University. She attained a BSc in Urban and Regional Planning at the University College of Lands and Architectural Studies (UCLAS), an MSc in Urban and Regional Planning and Management at UCLAS in Dar Es Salaam and Dortmund University in Germany. She also has a Licentiate degree in Infrastructure, Planning and Plan Implementation at KTH Royal Institute of Technology in Stockholm and earned her Ph.D. in Urban Planning Management from Ardhi University on Displacement and Resettlement of Communities. She has experience as a lecturer and has done research on issues such as outreach in resettlement, conservation, gender, regional development, small towns, and climate change. She is a registered town planner and a board member of the Trade Policy Review Board (TPRB) of Tanzania.

Ramadhani Hashim Matingas is a tutorial assistant at the School of Spatial Planning and Social Sciences at Ardhi University. He is also an MSc student at the Institute of Resource Assessment of the University of Dar es Salaam (UDSM). He holds a BSc in Regional Development Planning from Ardhi University. His research interests focus on climate change, natural resources, ecosystem management, spatial planning, rural development, and land tenure. Matingas has participated in several community outreach and consultancy services through Ardhi University. He is currently participating in a community outreach project aiming to improve land tenure security for more than 40,000 residents in the Mbarali District.

Part III
Participatory and Multi-Level Governance Approach toward Current Urban Challenges

Chapter 9
Qualities of Urban Planning and the Conflict Between Participatory Planning and Planning Standards: Evidence from Ethiopia

Behailu Melesse Digafe, Achamyeleh Gashu Adam, Gebeyehu Belay Shibeshi, and Mengiste Abate Meshesha

Abstract The Ethiopian hybrid planning system applies both top-down and bottom-up planning approaches simultaneously. This causes vague quality measurement indices of the urban plan and is a major obstacle for both the planning team and other stakeholders to measure quality. The chapter examines and dialectically discusses the contradictory measurement indices regarding the quality by taking Bahir Dar City Structural Plan Project as a case study. Both primary and secondary data were collected from the planning team and stakeholders for the study. This chapter argues that challenge arises from the system that uses two, often conflicting, yardsticks to measure quality, i.e. meeting the pre-defined standards and fulfilling the participants' interest. Therefore, it suggests that the quality of an urban plan should be primarily measured in terms of the local planning standard, which is the "public acceptance". Public acceptance here is described, measured and defined as the stakeholder's perception that the plan is of good enough quality for implementation.

Keywords Planning approaches · Quality measurements · Simultaneous application

9.1 Introduction

Today's urban areas require complex plans to balance all of the different elements that contribute to people's quality of life and the sustainability of their environment. The field of urban planning is as diverse as the communities themselves. Every urban plan uniquely addresses its community's challenges by coordinating the operation and development of the many different components of a city.

The driving force in Ethiopia for the creation of the urban-related policies and regulations has been the need to address the many existing urban development problems. Ethiopian urban development has been facing a range of problems that extend

B. M. Digafe (✉) · A. G. Adam · G. B. Shibeshi · M. A. Meshesha
Institute of Land Administration, Bahir Dar University, Bahir Dar, Ethiopia
e-mail: behailu.melese@bdu.edu.et

G. Alem Gebregiorgis et al. (eds.), *Planning Cities in Africa*, The Urban Book Series,
https://doi.org/10.1007/978-3-031-06550-7_9

across a wide area of concern: lack of good governance and inadequate public participation (Deepa Narayan et al. 2000), expansion of urban areas and sprawl (Terfa et al. 2019) together with its adverse effects; on public health (Frumkin 2002), on native ecosystems (Pauchard et al. 2006), on farmland, on infrastructure and public service costs, on energy efficiency and air quality, on travel and congestion, on industrial activities and infrastructural facilities (Shiferaw 1998); its psychic costs (Ewing 2008), unbalanced distribution of urban population, high level of unemployment, urban poverty and slum habitation (Kassahun and Tiwari 2012), and so forth. Therefore, Ethiopia has attempted to deal with these urban development issues through various planning mechanisms.

9.1.1 Ethiopian Urban Development Policy Framework

A planning system is typically driven by national policies and related regulations. All of Ethiopia's urban-related policies, regulations, and programmes characterise its planning system as a hybrid form that seeks to optimise both "bottom-up" or "discretionary" or "participatory planning" as well as "top-down" or "regulatory" or "planning standards". It involves simultaneously applying and expecting the best results from both top-down and bottom-up planning approaches. Firstly, the top-down approach must meet the pre-defined planning standards. Planning standards are formulated or endorsed by planning agencies of regional and/or federal governments to achieve the required quality of the urban plan and to meet long-term objectives. They can be in terms of locational, and/or space standards. Secondly, the bottom-up approach aims to attain residents' and stakeholders' satisfaction through participatory planning. Therefore, public participation is necessary to recognise and integrate the views of every stakeholder who has an interest, a voice, and a choice as it is believed that participatory urban planning increases the quality, legitimacy, and overall social, economic and environmental efficiency of planned development. However, there is often a conflict between the application of both top-down and bottom-up approaches when both are applied simultaneously and best results are expected from each of them in measuring the quality of an urban plan.

Consequently, the hybrid planning approach, which was intended to bring about a better quality of the urban plan, has become a major obstacle especially for planning participants and stakeholders to measure the quality of their local plan and achieve a better quality urban plan in Ethiopia. The aim of this study is to examine the conflicting quality measurements of the planning participants by taking the Bahir Dar City Structural Plan Project as a case study.

9.1.2 Urban Planning Approaches

According to Patchy Healey (1992), two main tendencies have marked the history of urban planning over the past five to six decades. On the one hand, there has been a tendency towards centralised decision-making, as well as increasing the role and power of technical experts. On the other hand, there have been demands for more participation in decision-making (Healey 1992). These two tendencies, very much conflicting with one another, have been labelled as the top-down and bottom-up approaches to planning (Murray et al. 2009).

9.1.2.1 The Top-Down Approach and Planning Standards

The essence of the top-down approach of planning is well illustrated by Patsy Healey, Glen McDougall, and Michael Thomas (Taylor 1998). According to these scholars, the process of application of planning standards involves five steps: the systematic analysis and definition of the problems, the identification of goals, the logical production of alternative plans/policies, the evaluation of the alternative plans/policies, and the implementation and monitoring of the chosen plan. The rational planning procedure embraces the use of planning standards. Faludi also emphasises the use of planning standards in the core of the rational planning process (Faludi 1973).

Planning standards exist for all urban uses, and they can either describe a projection of their growth at a specific time in the future or set certain limitations for their growth or their location. Planning standards can be broadly classified into locational standards and space standards (Olujimi 2009).

Locational standards are guidelines for assigning uses or facilities to land. They are presented in the form of the appropriate or ideal positioning of uses of land for the basic interaction needs of the users. In the process of development of locational standards, attention is given to safety from danger; nearness or remoteness of one land use from another in time and distance; compatibility and the social implications of the uses to the nearby community; land values and site development cost, etc.

Space standards are sets of planning standards that show the amount (or the extent) of space required to accommodate certain facilities, infrastructure or use. Space standards are defined frameworks by which all space should be allocated to improve space efficiency. They may be in the form of a unit of areas, estimated persons or other features per unit area, and are usually indicated in the form of minimum or maximum standards. Developers are not to go below the minimum standards, and the maximum standards are the upper limit. Space standards help to achieve proper use of land by preventing overcrowding and under-utilisation, and ensuring efficient functioning of various uses, facilities and services.

The use of planning standards has expanded internationally and they now comprise part of the planning practice in most countries (UN-Habitat 1999; Hooper et al. 2018). In most Western European countries, land-use regulations are legally binding and there has been substantial involvement of central government departments in standard development. However, the flexible and discretionary nature of English land-use planning is not binding, although the standards are used by planners (Gielen and Tasan-Kok 2010; Oxley et al. 2009).

In the Ethiopian urban planning system, the term "standard" is defined as the level and quality of site planning and zoning that conforms with established land-use policies and other regulations that are important to make urban areas comfortable and operational (MUDHC[1] 2014). In 2005, planning standards were formulated at the federal level to standardise urban plan and function across urban centres of the country. However, the implementation of these planning standards at the city level has not always been carried out as required. For example, Fetene et al. (2014) have reported that six out of seven broad land-use categories of Bahir Dar and Hawassa cities did not follow the necessary percentage proportions as per the standard for the respective land uses, and deviated by at least 25%, and at most 75%.

9.1.2.2 The Bottom-Up Approaches and Participatory Planning

The bottom-up approach in the planning process involves public participation to attain customers' satisfaction (Healey 1997). The World Bank (1996) defines public participation as "a process through which stakeholders influence and share control over development, initiatives, and the decisions and resources which affect them". The Ethiopian participation manual for urban planning also defines public participation as:

> any process that strives to inform, gather input or involve the public regarding decision-making processes. Public participation is the umbrella term to describe all levels of 'public' information, education, relations… input, involvement and collaboration. (MUDHC 2007)

The importance of public participation in the urban development processes is acknowledged by contemporary planning theories, advising that public participation creates the possibility to attain more sustainable results. Research in this field shows that citizen participation can generate trust, credibility and commitment regarding the implementation of policies (Van Empel 2008). Therefore, public participation in urban planning has become one of the well-known tools used to integrate the interests and requirements of all stakeholders. And it has also become a compulsory way to recognise and integrate the views of every stakeholder who has an interest, a voice and a choice, as it is believed that participatory urban planning improves the quality, legitimacy, and overall social, economic and environmental efficiency of planned development. Participatory planning helps to harmonise views among stakeholders in planning processes, prevents conflicts between parties representing various interests and contributes to a long-term consensus. Implementation-oriented and sustainable land use planning processes need to be participatory, and to involve the urban population which manages the land and natural resources (Betke 1999). According to CNPPAM[2] Benchmarking and Best Practice Program, one benefit of

---

[1] Ethiopian Ministry of Urban Development, Housing and Construction.

[2] Committee on National Parks and Protected Area Management.

public participation is coming to understand the customers' expectations and needs (PWCNT[3] 2002).

According to the International Organization for Standardization (ISO), customers should also be involved even in the technical standards development or improvement period. This is very fundamental to ensure that technical standards for the quality of service are as comprehensive as possible and correspond to actual consumer needs. Public participation in standardisation is not only important for standard development but it can also play a crucial role in raising public awareness of the existence of agreed standards. Thereby, it helps the customer to demand the rendered service according to these standards (ISO 2001).

A vast range of public participation is possible with different aims and characteristics. The International Association of Public Participation (IAP2) (2014) has developed the Spectrum of Public Participation (SPP) to make more comprehendible the role of the public (end users) in planning and decision-making, and the level of power the community or the public has over planning or decision-making. The tool has long been used in different community involvement projects and has remarkable benefits that have been proven by its efficiency for many years in this field. As many practitioners and organisations find the Spectrum very helpful, the IAP2 claims the Spectrum is "quickly becoming an international standard". The Spectrum identifies five levels of public participation (or community engagement). They are described on the Spectrum ranging from no influence (Inform) to total influence (Empower) (IAP2 2014). In Ethiopia, the public participation process may work on the following four levels:

- Information sharing is one-way communication often involving disseminating information about an intended development project, programme, or strategy.
- Consultation is about gaining stakeholder input on proposed activities.
- Collaborative decision-making means engaging groups to decide jointly about development activities and resources that affect them.
- Empowerment is a deeper level of participation, where beneficiaries and other key groups initiate action and take control over development decisions and resources.

9.1.2.3 The Hybrid Planning Approach

As there are many weaknesses arising from both planning standard (regulatory) and participatory planning (discretionary or flexible) approaches, countries such as France, the United Kingdom, China, Singapore and Australia have introduced a hybrid planning approach to avoid weaknesses of these two approaches which helps to minimise the degree of certainty and to increase the degree of flexibility of the regulatory approach (Omer 2017; Elliott 2008; Carmona et al. 2003). However, there have not been adequate studies available that show the effectiveness of a hybrid planning approach, especially from the point of view of measuring plan quality and its performance of achievement.

[3] Parks and Wildlife Commission of the Northern Territory.

9.1.3 Quality of the Urban Plan

In Ethiopia, concerned government bodies have exerted much effort into achieving the quality of the urban plan in alignment with the existing planning policies. However, different studies continue to report on the existence of the problem of urban plan quality in Ethiopia. One of the most commonly mentioned factors that affects the quality of the urban plan is related to the basic map of the urban areas, which is an important input for the preparation of an urban plan Digafe and Feleke (2018). The other common factors are:

- replication of experiences and planning standards from other countries with widely different settings,
- obsession with land use and physical infrastructures with little attention to the actual needs of the population, and
- the preparation of urban plan with little or no involvement of the public, being often mentioned as the main and sometimes the only factor (MUDHC 2007).

Quality denotes excellence in service and products, especially to the degree that they conform to standard requirements and satisfy customers' wants. Therefore, quality measurements of services, including urban planning, can be seen from two perspectives: the degree that which they conform to the standard planning requirements and the degree that they satisfy customers' wants. The internal perspective is defined as a zero error rate, zero defect (Hartnett et al. 1988) i.e. the ability to meet all the pre-defined standard requirements (Parasuraman et al. 1985). From this perspective of quality, an urban plan can be evaluated using the pre-defined planning standard, which constitutes the basic features of urban analysis. Whereas, the other perspective sees service quality in terms of customer perception, customer expectation, customer satisfaction, customer attitude and customer delight (Sachdev and Verma 2004).

Hence, the quality measurement of an urban plan involves either or both perspectives of quality that depend on planning approaches. That means, for the top-down planning approach, the degree to the output of an urban plan conforms to the pre-defined planning standards and requirements. For the bottom-up planning approach, one can say that an urban plan is of good quality when it complies with the requirements specified by the local communities and stakeholders who have participated in the project.

However, there continues to be ongoing criticism as to whether channels of participation in the development process are truly in the spirit of the process or merely manipulation, which planning teams attempt in the development and even the collaborative process (Arnstein 1969; Ansell and Gash 2007). Without redistribution of power, participation can be an empty and frustrating process for those who are powerless, particularly for ordinary citizens. Therefore, this chapter will study the acceptance of local plans by participants using the 2020–2030 Bahir Dar City Structural Plan Project as a case study from the point of view of the "empower" level of public participation.

As mentioned, the goal of the "empower" level of public participation is to place final decision-making in the hands of the public. Thus, at this level, the customers or stakeholders are considered as being satisfied when all things will be implemented just as they want.

9.2 Methodology

The study takes the 2020–2030 Bahir Dar City Structural Plan Project as a case study. Both primary data and secondary data were collected from the planning team and stakeholders' representatives for the study. The primary data were collected to identify quality measurement indices of the two parties through questionnaire and focus group discussion. The key stakeholders that participated were:

- Community (Ketena[4] level) representatives,
- Officials, politicians, professionals and staff members of concerned regional institutions and the city administration,
- Elders, knowledgeable persons, and professionals from the community,
- Investors, developers, and private firms, and
- NGOs, Community-Based Organisations (CBOs), social organisations and associations.

Secondary data were also collected mainly to analyse the system established to maintain the quality of the urban land-use plan. Therefore, documents of national policies, strategies, laws, regulations, manuals and standards were reviewed. In addition to these, planning reports of the project office were examined.

The collected data are dialectically discussed. The discussion focuses on only some relevant, nationally pre-defined planning standards which were challenged by the stakeholders during the preparation process of the 2020–2030 Bahir Dar City Structural Plan. Therefore, the Urban Population Density Planning Standard, the Urban Land-use Proportion Planning Standard, the Standards for Right of Way (ROW) and the Spacing of Streets in City Centres and Built-up Areas are the only planning standards examined under this study.

Aiming to address the difficulty of measuring the quality of an urban plan, this study attempts to answer the question: "how are the quality measurement indices of each approach contradicted?".

[4] Ketena is the smallest geographical hierarchy (lower than kebele) that is introduced informally for easement of administrative purpose in Bahir Dar by the local administrative body without appointed leaders. Normally a ketena can accommodate 2,000–3,000 people.

9.2.1 *Case Study—The 2020–2030 Bahir Dar City Structure Plan (SP)*

Bahir Dar is one of the nine regional capital cities of Ethiopia located at the geographic co-ordinates of 11° 38' N latitude and 37° 15' E longitude, with a population size of over 500,000 in 2020 (BDCSPPPO 2020). Based on the planning law of the country, Bahir Dar Structure Plan Preparation Project (BDSPPP) office was established and staffed with over 90 professionals who had different educational backgrounds and fields of specialisation for the revision work of the city plan. The 2020–2030 Bahir Dar City SP preparation process used two major approaches to achieve the qualities of the outputs of the city plan: participatory planning (bottom-up approach) and planning standards (top-down approach). The simultaneous application of these approaches resulted from the Urban Planning Proclamation 574/2008, which states in its preamble:

> It is vital to create a favourable and an enabling condition for public and private stakeholders to fully participate in the process of urban plan initiation, preparation and implementation on the basis of national standards. (FDRE[5] 2008a)

On the one hand, the urban planning proclamation 574/2008 of article 5 number 5 makes participatory planning an obligatory method in Ethiopia to ensure the satisfaction of the needs of society through public participation. Furthermore, article 15 of the proclamation, states that public hearings are mandatory before the approval of plans. As a result, the 2020–2030 Bahir Dar City SP preparation process has tried to be transparent and adequately communicate with the public at large, public institutions and the kebele[6] councils, and take relevant suggestions and objections as inputs.

On the other hand, as stated in the Urban Planning Proclamation 574/2008 article 5 number 2, national standards must be maintained in any urban plan preparation project in the country, because they are considered vital to:

- bring about coordinated, efficient, transparent urban plan preparation and implementation throughout the country,
- help prepare uniform plans in urban areas,
- serve as a guide for building a better quality of urban areas,
- facilitate and ensure the planning, design and implementation of good coverage, sustainable, well served, resilient, comfortable, green and beautiful urban environment,
- address local practical issues facing professionals engaged in urban planning, design, construction and improvement,
- implement government development policies, strategies and programmes within this understanding to bring about sustained developments, and thereby realise economic development goals in urban centres of the country (FDRE 2008a).

---

[5] Federal Democratic Republic of Ethiopia.

[6] *Kebele*: the lowest hierarchical level of governmental administrative body in Bahir Dar City. Normally a Kebele can accommodate population of 5,000–10,000.

Consequently, meeting the pre-defined planning standards becomes well-thought-out as some criteria for the good quality of a plan by the in-charge governmental bodies is to follow up and supervise the 2020–2030 Bahir Dar City SP preparation process. Therefore, as a major method of analysis, they have required urban areas to compare the facts and figures of their existing situation with the pre-defined standards set at national and regional levels. However, there are many contradictions even among the planning standards: on the same subjects issued in a different year of publication by the same planning agencies and/or on the same subjects issued in the same year of publication by different planning agencies.

However, the differentiation and contradictions of these Ethiopian urban planning standards are not within the scope of this chapter. For the case study only three recently issued national planning standards are purposefully selected to examine the conflicting yardsticks used by those involved in the preparation process to measure the quality of urban plans.

9.3 Results and Discussion

9.3.1 Participatory Planning and the 2020–2030 Bahir Dar City SP

The 2020–2030 Bahir Dar City SP preparation project office tried to identify and involve the key stakeholders based on the structure plan preparation manual developed by the Federal Urban Planning Institute (MWUD[7] 2006). Table 9.1 shows the identified key stakeholders who have participated in the project, including their potential interests. According to the collected data from the project office, the decision-making role is only given to the Community (Ketena level) representatives. This means that a planning standard can never be applied, and neither can the structure plan ever be approved for implementation unless the ketena representatives give their consent.

9.3.2 Planning Standards and the 2020–2030 Bahir Dar City SP

Data from the project office shows that the planning team was trying to apply all urban planning standards to the plan for the ketena representatives to give their consent. The following three tables demonstrate those which are selected for this study: The Urban Population Density Planning Standard, the Urban Land-use Proportion Planning Standard, and ROW and Spacing of Streets in City Centres and Built-up Areas.

[7] Ministry of Works and Urban Development.

Table 9.1 Key stakeholders and their potential interests

Key stakeholders	Potential interests
Community (Ketena level) representatives	Provision of alternative strategies and solutions
	Reaching consensus and decision-making
Professionals and Staff Members of: Regional Urban Planning Institutions, The City Administration	Identification of planning issues, cause-effect relation of problems, and opportunities and constraints
	Maintenance of pre-defined planning standards
	Facilitation of the planning process to reach consensus among stakeholders and thereby to address all the identified planning issues
Officials of Sector offices—and institutions, and politicians	Ensuring the proper public participation
	Identification of problems and city potentials
	Identification of government and political interests
	Harmonising with different local interests
	Provision of alternative strategies and solutions
Knowledgeable Persons such as professionals and elders	Provision of relevant information
	Identification of city's basic problems and potentials
	Formulation of strategies to address the identified issues
	Preservation of culture, history, heritage, and identities of the city
Private Firms such as investors and developers	Identification of development constraints
	Provision of alternative strategies and solutions
	Creation of favourable conditions for investment and development
Social—Organisations and Associations such as NGOs and CBOs	Identification of social problems, development constraints, and other basic problems and potentials of the city
	Provision of alternative strategies and solutions
	Facilitating the proper public participation Identification of social problems, development constraints, and other basic problems and potentials of the city

Source Adopted from the structure plan preparation manual (MWUD 2006)

They, including all other pre-defined planning standards set at national and regional levels, were required to be implemented in the structural plan by governmental planning agencies (professionals and staff of relevant urban institutions, city administration, regional bureaus and institutions). However, all these parties, including the planning team, were challenged by other stakeholders, including the Ketena level representatives, not to apply the standards to the plan.

9.3.2.1 The Urban Population Density Planning Standard

The plan of Bahir Dar, which was prepared to serve for 50 years (1965–2015) proposed a population density (total population per total built-up area) of 1,393 inhabitants per hectare. However, the plan was revised in 1996 before it reached the end of its planning period. The plan (1996–2006) proposed a population density of 122 inhabitants per hectare. The following (2006–2016) plan proposed a population density for Bahir Dar of 1,402 inhabitants per hectare. However, in fact, the population density of Bahir Dar in 2006, was 91 inhabitants per hectare (MWUD 2006). By 2017 the population size of the city of Bahir Dar was 362,290 (BDCSPPPO 2019) and the total existing built up area was 6,348.49 hectares (BDCSPPPO 2017), which means the population density then was 57 inhabitants per hectare. The 2020–2030 Bahir Dar City SP was supposed to be prepared according to the national population density standard, as it has a direct impact on the total amount of spatial expansion of the city.

Table 9.2 shows the urban population density standard set by the Ethiopian Ministry of Urban Development and Construction in 2012 for all urban areas of the country. This density standard is the first selected standard among many other urban planning standards that the planning team of Bahir Dar and different governmental planning agencies tried to apply to the plan. As can be seen in the table, the population size of urban areas at the end of their planning period has to be above 1 million to have an average population density of 500 inhabitants per hectare (MUDHC 2012).

According to the information collected from both the planning team and the secondary data, the major justifications for this standard are similar to those in the reviewed literature. The standard addresses the excessive spatial growth of cities and helps to prevent urban sprawl as it is recognised as having a detrimental effect

Table 9.2 The urban population density planning standard

Range of total population size	Average population density standards
2,001–20,000	100 in/ha
20,001–50,000	200 in/ha
50,001–100,000	300 in/ha
100,001–1,000,000	400 in/ha
Above 1,000,000	500 in/ha

Source MUDHC (2014)

on sustainable development (Brueckner 2000). Furthermore, numerous researchers agree on most of the interrelated adverse effects of urban sprawl—on public health (Frumkin 2002); on native ecosystems (Pauchard et al. 2006); on farmland, on infrastructure, and public service costs, on energy efficiency, and air quality, travel and congestion and psychic costs (Ewing 2008). Based on the above-mentioned density standard of the country and the current population size of the city, the population density of Bahir Dar is proposed by the planning team to be at least 400 inhabitants per hectare within the coming ten years of the planning period.

However, according to the primary data collected from the key stakeholders that participated, they could not reach an agreement on the density standard. One of their arguments against maintaining this density standard hinged on the existing lack of financial capacity of both the city administration and the inhabitants. On the one hand, the cost of compensation, relocation and resettlement of the many already existing low-density built structures would make maintaining the standard unaffordable to the city administration. On the other hand, the citizens themselves do not have the financial capacity to build multi-storey buildings that can accommodate the required number of inhabitants within the specified limited space. The other argument against maintaining the density standard that was presented by the stakeholders was the preference of most people for low-density housing and the demand to have a large plot of land with a detached house and a garden that is in touch with nature, reserved lifestyles, together with the existing low awareness about how to live in multi-story buildings.

9.3.2.2 The Urban Land-Use Proportion Planning Standard

Table 9.3 shows the Urban Land-use Proportion Planning Standard set by the MUDHC in 2014 for all urban areas of the country. This land-use proportion standard is the second selected standard among many other urban planning standards that the

Table 9.3 The Urban Land-Use Proportion Planning Standard

Proportion	Land-use types to be included
30%	Road and related infrastructure within a road right of way including pedestrian ways and crossings, vehicular ways (major, collector and local), bikeways, green areas along (side and in the middle) roadways, utility lines (water, drainage, electricity, water, etc.) along with roadways, road junctions and roundabout areas, car parking and bus stop areas, street market areas;
40–45%	Built-up (Building) areas for Housing, working, and production, worshiping, etc.
25–30%	Natural environment Parks, garden/agriculture, and green areas (publicly and privately owned), recreational areas, playgrounds, urban forest, wetland areas, grass and bush areas, quarry (rock and other minerals) sites, river and stream areas, sports areas, cemeteries and open worshiping areas, open market areas, plazas and squares

Source MUDHC (2014)

planning team of Bahir Dar and different governmental planning agencies have tried to apply to the plan. In accordance with the standard, every urban area in Ethiopia should have either a land-use proportion of 30%, 40%, 30% if they are comparatively new with few historic features and developments, or a land-use proportion of 30%, 45%, 25% if they are comparatively old with many historic features and developments (MUDHC 2014).

According to the information collected from the planning team and secondary data, the rationale behind this standard is the aim to address the existing urban development problems of the country and to balance the important development types (road and related infrastructure, built-up areas, and natural environment) through which, they believe, sustainable development can be brought about.

However, the key participatory stakeholders could not agree with the land-use proportion standard. As reported by them, the first justification was the difficulty of applicability. As they stated, it is very difficult to measure the amount of land dedicated to the "natural environment" land-use type, separate from the built-up (building) areas in the case study area, and reserve accurately the required proportion; to measure the small-scale green areas, the structure plan would require detailed survey works that demand a great deal of time, money, and specialised human resources, which is directly incompatible with structure plan but similar to the lower level and detailed type of urban plan (i.e. Neighbourhood Development Plan). For this reason, the standard could not apply in the 2020–2030 Bahir Dar City SP project.

The second justification was related to the financial and technical capacities of the city. The existence of built structures would result in compensation costs, which exacerbate the shortage of finance. Poor enforcement capacity of the city was also presented as justification for not agreeing with the application of the land-use proportion standard. In short, they were advocating a strategy that the standard should apply only to prevent the city from emerging related problems in the future rather than cure the city of the already existing related problems. Otherwise, insensitive universal application of the standard could bring about additional adverse effects on both the inhabitants and the city administration.

The third justification was related to a mismatch with the context. According to them, the city is already home to two major, large, water bodies, namely Lake Tana and Abay (Nile) River. Together with their tributaries and the large space of the green area, 71% of the total planning area is currently the natural environment (environmental sensitive area); green buffer areas, wetlands, mountain green areas and urban agriculture, resulting in a land-use proportion that exceeds the maximum limit of the standard. Consequently, the stakeholders were not willing to reduce the natural environment as per the standard, and due to this, it could not be applied in the 2020–2030 Bahir Dar City SP project. For better insight see the land-use proposal of the 2020–2030 Bahir Dar City SP (Fig. 9.1—natural environment areas are coloured light green and water bodies are coloured light blue)

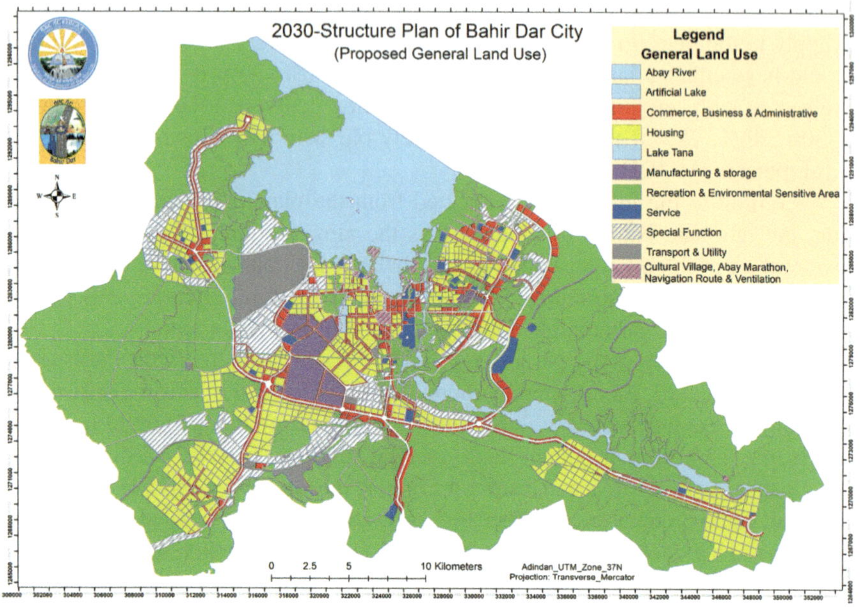

Fig. 9.1 The proposed general land use of the 2020–2030 Structural Plan of Bahir Dar City (*Source* BDCSPPPO 2020)

Table 9.4 The standards for right of way (ROW) and spacing of streets in city centres and built-up areas

Category of streets	ROW (street width)	Spacing
Principal arterial street (PAS)	30–60 m	1,000–2,000 m
Sub-arterial street (SAS)	20–60 m	500–1,000 m
Collector street (CS)	15–20 m	250–500 m
Local street (LS)	6–12 m	30–250 m

Source MUDHC (2017)

9.3.2.3 The Standards for Right of Way (ROW) and Spacing of Streets in City Centres and Built-Up Areas

Table 9.4 shows the Standards for Right of Way (ROW) and Spacing of Streets in City Centres and Built-Up Areas set by the Ethiopian MUDHC in 2017. This street standard is the third selected standard among many other urban planning standards that the planning team of Bahir Dar and different governmental planning agencies have tried to apply to the plan. Based on the standard, the spacing between two similar categories of streets cannot be below or beyond the range shown in Table 9.4.

According to the information collected from the planning team and secondary data, the standards have many objectives. These are: to improve the quality of existing streets in urban areas of the country; facilitate and ensure the planning, design and implementation of green coverage, and sustainable, well served, resilient, comfortable, green and beautiful streets; implement street designs uniformly in urban areas of Ethiopia; address local practical issues facing professionals engaged in street planning, design, construction and improvement; help implementers, actors, agencies, institutions, community, private and public organisations to understand the significance, challenges, design criteria and standards, geometric designs, applicability, and sustainability opportunities of street planning and design; implement government development policies, strategies and programs; bring about sustained developments and thereby realise economic development goals in urban centres.

The 2020–2030 Bahir Dar City SP proposed a superblock with major land use and street network (Fig. 9.2). The superblock is defined by major streets: two 15 m wide Collector Streets (CSs), a 30 m wide Sub Arterial Street (SAS), and a 40 m wide Principal Street which is subdivided by many small blocks with 10 m wide Local Streets (LS). But its size is about 660 m × 770 m exceeding the street spacing standard meaning that at least two CSs that connect the two opposite sides should be introduced into the superblock as per the spacing standard demand. However, the participating stakeholders of the project could not accept this standard and challenged the application of the standard in the structural plan of Bahir Dar with the following arguments. According to them, meeting the national planning standard or reserving appropriate width (right of way) for streets in the SP would

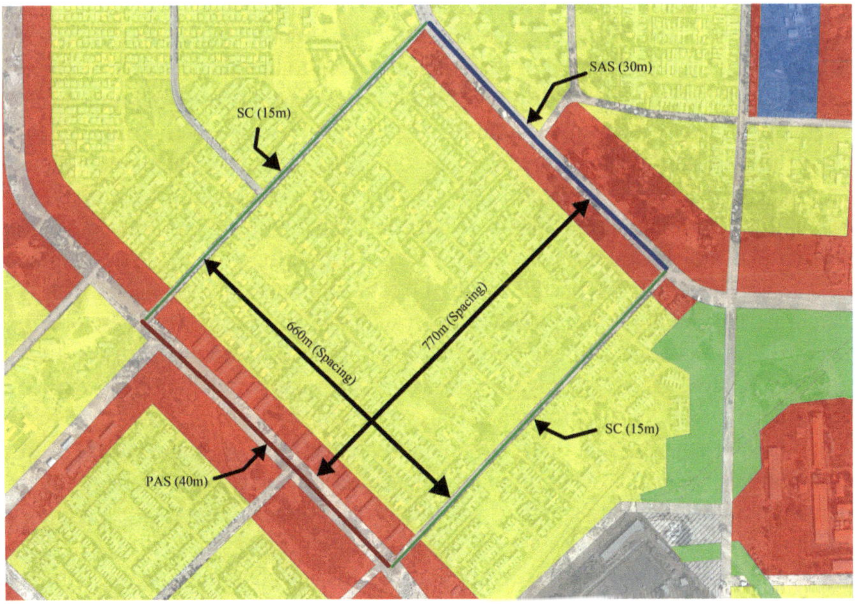

Fig. 9.2 A zoomed-in image of a super block of the 2020–2030 Structural Plan of Bahir Dar City: proposed major land use (major road network) (*Source* BDCSPPPO 2020)

be difficult because of the existence of built structures and environmentally sensitive areas, shortage of financial resources for compensation, relocation, and resettlement from existing developments, and low technical capacity for enforcing and managing the urban plans and designs. Moreover, the existing large coverage of wetlands, flood areas, fertile ground, swampy areas, lakeshore areas, riverbank areas, and streams, are required to be protected for environmental benefits. Yet streets demand a huge amount of finance as they need extra structure to make them stable or extra length to get more suitable space, making it difficult to implement the standards. For this reason, the standard could not be applied in the 2020–2030 Bahir Dar City SP project. (For further insight, see Fig. 9.1: The Proposed Land Use of the 2020–2030 Structural Plan of Bahir Dar City).

9.3.3 The Conflict Between Participatory Planning and Planning Standards During Bahir Dar City SP Preparation

The study reveals that the hybrid planning approach created conflicts between two groups of stakeholders. The first group of stakeholders included the planning team and staffs of different governmental planning related agencies, such as city administration, regional bureaus and institutes, while the second group was comprised of stakeholders from the general public, NGOs, CBOs, the private sector and the like (see Table 9.1). The first group advocated the top-down planning approach, which is in favour of the application of planning standards to the plan. Their major justifications are based on existing urban planning law and widely used imported criteria. Consequently, they measure the quality of a plan by the degree of conformance to the pre-defined planning standards. Thus, according to these proponents of standards, an urban plan that does not meet these standards is considered poor quality. As a result, they rejected interests that were not compatible with the pre-defined planning standard, as they believed the interests were not useful for sustainable development.

The second group advocated for the bottom-up planning approach, which is in favour of stakeholders' participation. Their major justifications were based on their legal right to make an ultimate decision on any local development activity and the practical context. Consequently, they measured the quality of a plan by its degree of acceptance by the local community. Thus, according to these proponents of participation, an urban plan that does not gain acceptance from the stakeholders or local communities is considered poor quality. In this case, they rejected some of the pre-defined national planning standards, as they believed the standards were not useful for their context.

In other words, the study reveals that the hybrid planning approach initiates disagreements among planning participants with conflicting ideas: because it allows for the simultaneous application of two, often conflicting yardsticks used by those involved in the process to measure the quality of urban plans, i.e. meeting the pre-defined standards and fulfilling the participants' interest. Therefore, it exacerbates dispute rather than serving as a platform for resolution.

Standards are criticised by the proponents of participation as they limit local ideas and ways of thinking; underestimate the preferences of local communities; fail to recognise their own specific context and urban development dynamism, and attempt to treat Bahir Dar as having uniform needs and expectations with other different urban centres. Therefore, these stakeholders see the standards as adversely affecting the city's features which have existed over time; a factor in losing the city's own identity (Oktay 2002); a force abolishing the decision-making power of local communities and replacing them with the knowledge of experts who have not been involved in the structure plan preparation project. Moreover, these standards have also been viewed as inappropriate and inefficient by the stakeholders because they bring about many unwelcome consequences, such as leading to huge distractions from existing development and cumbersome technical and administrative procedures.

Participation at "empower level" was criticised by proponents of standards as most of the participatory residents were found unqualified to make good decisions in urban development activities compared to good practices of participatory planning in other areas. For example, some participants expressed a personal interest to be included in the planning. However, this was very difficult for the professionals to fulfil because of the resources and capacity the city has. This alone can lead to disagreement among the participants. Therefore, some of the political elite and participating planners argued that such people were not well educated enough to determine their own fate. Thus, the bottom-up approach through public participation is not welcomed by the elite (Dimitrova 2014).

Moreover, the planning approach of the country was criticised by proponents of the standard for deconstructing any well-established planning system without replacing it with a better or, at least, another defined method of urban intervention.

For the proponents of standards, the strong attachment to planning standards over time would be an adequate reason to keep them in use. However, this was not a good reason for the opponents to accept the usefulness of the standards, because they are too rigid to accommodate the emergence of new and better tools or methods of planning practice. Proponents of standards were also arguing that standards are proposed with a range to which the characteristic studied should be conformed and this elasticity allows freedom of choice in planning. However, for the opponents, this value range still imposed certain limits and it could not accommodate the needs of the planning participants. Thus, it reduced the planning possibilities of the city's community (Healey 1997).

9.4 Conclusion and Recommendations

The hybrid planning approach of the country becomes a major obstacle to planning participants reaching a consensus regarding the quality of their local plan; because, according to the approach, the quality of the plan is gauged using two contradictory measurement indices, i.e. meeting the pre-defined standards and fulfilling the participants' interest. Since this chapter is being considered an introduction to the dialectical debate, it describes the conflicts between the measurement indices used by the planning participants of the Bahir Dar city plan, the case study.

Although standards should normally be developed based on the consensus of different concerned parties (Xie et al. 2015), the hybrid approach does not allow the local communities, to whom the standards are applied, to be involved in standards development. Therefore, the standards being applied become a source of conflict as they are imposed on those parties without their consent. Hence, the study confirms that standards cannot be as efficient as expected if their development process follows a top-down approach rather than a bottom-up one (Aim 2021).

These contradictory measurement indices of the hybrid approach threw the planning participants into confusion. And the quality attributes became vague and difficult to measure. The study identifies two major issues that cause confusion and vagueness regarding the quality measurement indices of the urban plan in Ethiopia.

Firstly, the hybrid approach considers the urban communities as different conflicting actors in a single urban planning project. Therefore, the relationship of the local communities to the planning authority is simultaneously as clients, customers and stakeholders. On the one hand, as clients or customers, they are meant to be part of the ultimate decision-making body regarding the quality of the plan. Here, they may not have the authority to develop or change the pre-defined planning standards. However, they should express their satisfaction with the service outcomes and components of the plan through different mechanisms, such as final approval of the plan for implementation. On the other hand, as stakeholders, they are intended to be part of the entire plan-making process. Here, the local community can even have the authority to develop or change the planning standards, which totally contradicts their formerly mentioned role. Therefore, they, finally, may not be needed to approve the plan for implementation, as they have already been involved in the entire process. Such conflicting roles of the local community create confusion regarding urban plan quality indices.

Secondly, the hybrid approach gives the planners conflicting roles during the plan preparation process. On the one hand, planners must play their role as the government agent, to serve policymakers and the administrative bodies. Here, they are responsible for maintaining the pre-determined planning standards. On the other hand, they must also play their role as local community agents, to serve the urban community. Here, they are responsible for communicating and protecting the interests of the local community. Therefore, if the local community has a problem with pre-determined planning standards, they must see what to change and adapt to the local context, which conflicts completely with their formerly mentioned role that is simply to maintain the standards. Such conflicting roles of the planners create confusion regarding the urban plan quality indices.

Aiming to address these aforementioned issues, the study forwarded the following possible mechanisms. Clearly, it becomes impossible to be productive if individual planning parties work only towards their own interests (Fewster-Thuente 2011). Furthermore, the hybrid planning approach could not help the urban communities to agree on common goals that would unite them and encourage them to strive for clear results. Thus, the study confirms that collaborative activities without a common goal can never be successful. Therefore, helping the local community to agree on common planning goals should be the most important role of an urban planning system.

The study recommends that the common goal in urban planning should be primarily to benefit the local community, but not others (Rodela and Udovč 2008; Sachdev and Verma 2004). And others cannot determine benefits without the consent of the beneficiaries themselves. Therefore, the quality of an urban plan should be measured primarily in terms of whether the local community is satisfied with the decisions made regarding the plan.

Satisfaction should be measured by the acceptance of the ultimate decisions by the beneficiaries or the local community (Hilchey and Hurych 1985). Here, community acceptance is only concerned with the participants' (stakeholders' representatives') attitude and opinion towards believing in the decision. And it is measured as the stakeholders' perception of the decision to be acceptable (approved) or not acceptable (rejected) for implementation. This is because, according to the planning law of the country, any urban plan must be approved by the local administrative council which is made up of representatives of the local communities (FDRE 2008b).

As a result, the roles of all planning participants would be clear and no longer confusing, and centrally developed planning standards would not be considered as criteria to measure the quality of a local plan. Instead, they would serve merely as a guideline for a local planning process. Meaning, they should not be too detailed so as not to have the power to override local common or majority interest. In other words, the centrally developed standards should be advisory but not obligatory or should set only the basic elements to create room for open choices and more options for local interests in dealing with regulations.

In short, the participatory planning approach with little modifications should be, therefore, the dominant approach that should replace the hybrid planning approach of the country. As a result, the ultimate decision-making as to the quality of an urban plan would be given only to the local community. To address the aforementioned major issues, obligatory but regularly updatable local detailed standards should be developed for every urban centre uniquely; because standards that are centrally developed are often time-consuming and then become rigid. Therefore, they should not be developed at the central government level (at large scale), but at the immediate higher local government level (at small scale). This scale can allow room for the required level of participation (empower) with the local community. Local communities are to be mainly those that can directly affect the decisions and can be affected by them (MUDHC 2007). Therefore, the local obligatory standards would consider the capacity of every urban centre that is supposed to be responsible for implementing the standards in terms of technical, financial, and other resources' capacities.

Updating and revision work on obligatory local standards should at least be done at the beginning of every planning period of an urban plan to incorporate each urban centre's recent and dynamic conditions and needs. In addition, all urban plan preparation and revision project work should provide feedback on these standards. The feedback should show how the standards are compatible with the existing general context. It must also be provided in such a way so as to update it (if necessary) regularly to incorporate the unique and dynamic social, economic, and environmental requirements of the specific urban centre.

Finally, the local community should be well informed about the advantages of conforming to and disadvantages of deviating from the widely used common standards before they make a unique and final decision regarding the local (specific) obligatory standards. One of the major responsibilities of the planning professionals should be to make all of the participating bodies aware of the ideas and intentions behind the various planning standards that are widely used in other urban centres. Maintaining the commonly known planning standards should never be the responsibility of the planners. In conclusion, unless the local community accepts these standards as relevant to and useful for their specific cases, no standards should be applied in the name of wide acceptability.

References

Aim (2021) Understanding standards development. https://www.aimglobal.org/understanding-standards-development.html. Accessed 7 Sep 2021

Ansell C, Gash A (2007) Collaborative governance in theory and practice. J Public Adm Res Theor 1–29

Arnstein NS (1969) A ladder of citizen participation. J Am Inst Planners 30:216–224

BDCSPPPO (Bahir Dar City Structural Plan Preparation Project Office) (2017) Bahir Dar City Stractural Plan Preparation Project Office: Bahir Dar Integrated Development Plan (BDIDP)—Evaluation Report. Bahir Dar—Ethiopia

BDCSPPPO (Bahir Dar City Structural Plan Preparation Project Office) (2019) Bahir Dar City Stractural Plan Preparation Project Office: Population Projection Report: Bahir Dar 2017–2063. Bahir Dar—Ethiopia

BDCSPPPO (Bahir Dar City Structural Plan Preparation Project Office) (2020) Bahir Dar City Stractural Plan Preparation Project Office: Bahir Dar City Stractural Plan Report. Bahir Dar—Ethiopia

Betke D (1999) Land use planning: methods, strategies and tools

Brueckner JK (2000) Urban sprawl: diagnosis and remedies. Int Reg Sci Rev 23(2):160–171

Carmona M, Gallent N, Carmona S (2003) Delivering new homes: processes, planners and providers. Taylor & Francis, Abingdon, United Kingdom

Digafe BM, Feleke HB (2018) Expectations against pre-considerations on quality of urban land use plan: the case of amhara region-ethiopia. Afr J Land Policy Geospatial Sci 1(3):32–44. https://doi.org/10.48346/IMIST.PRSM/ajlp-gs.v1i3.13689

Dimitrova E (2014) The "sustainable development" concept in urban planning education: lessons learned on a Bulgarian path. J Clean Prod 62:120–127. https://doi.org/10.1016/j.jclepro.2013.06.021

Elliott D (2008) A better way to zone: ten principles to create more liveable cities. The University of Chicago Press

Van Empel C (2008) The effectiveness of community participation in planning and urban development. WIT Trans Ecol Environ 117:549–556

Ewing RH (2008) Characteristics, causes, and effects of sprawl: a literature review. Urban Ecol Int Perspect Interact between Humans Nat 21(2):519–535. https://doi.org/10.1007/978-0-387-73412-5_34

Faludi A (1973) The "systems view" and planning theory. Socioecon Plann Sci 7(1):67–77. https://doi.org/10.1016/0038-0121(73)90012-8
FDRE (Federal Democratic Republic of Ethiopia) (2008a) Urban planning proclamation. Addis Ababa, Ethiopia
FDRE (Federal Democratic Republic of Ethiopia) (2008b) Urban planning proclamation. Addis Ababa, Ethiopia
Fetene MW, Serbeh-Yiadom KC, Asfaw M (2014) A critical review of the application of planning standards and norms in the planning of Bahir Dar and Hawassa Cities, Ethiopia. Developing Country Stud 4(5):66–79
Fewster-Thuente L (2011) Working together toward a common goal: a grounded theory of nurse-physician collaboration. LOYOLA UNIVERSITY CHICAGO, 24
Frumkin H (2002) Urban sprawl and public health. Public Health Rep. https://doi.org/10.1016/S0033-3549(04)50155-3
Gielen DM, Tasan-Kok T (2010) Flexibility in planning and the consequences for public-value capturing in UK, Spain and the Netherlands. Eur Plan Stud 18(7). https://doi.org/10.1080/09654311003744191
Hartnett JL, Schechter SH, Kropp DH (1988) Managing quality: the strategic and competitive edge. Acad Manag Rev 13(4):656. https://doi.org/10.5465/amr.1988.4307572
Healey P (1992) Planning through debate: the communicative turn in planning theory. Town Plann Rev 62:143–162. https://doi.org/10.3828/tpr.63.2.422x602303814821
Healey P (1997) Collaborative planning: shaping places in fragmented societies. UBC Press, Vancouver. https://doi.org/10.1007/978-1-349-25538-2
Hilchey S, Hurych J (1985) User satisfaction or user acceptance? Statistical evaluation of an online reference service. RQ
Hooper P, Boruff B, Giles-Corti B (2018) Testing spatial measures of public open space planning standards with walking and physical activity health outcomes: findings from the Australian national liveability study. Landscape Urban Plann 171:57–67
IAP2 (International Association of Public Participation) (2014) IAP2's public participation spectrum, international association for public participation. https://cdn.ymaws.com/www.iap2.org/resource/resmgr/foundations_course/IAP2_P2_Spectrum_FINAL.pdf. Accessed 1 Aug 2020
ISO (International Organization for Standardization) (2001) ISO/IEC/GEN 01:2001mj/pc/12554061ISO/IEC Statement on consumer participation in standardization work. https://www.iso.org/files/live/sites/isoorg/files/archive/pdf/en/copolcoparticipation_2001.pdf. Accessed 29 July 2020
Kassahun S, Tiwari A (2012) Urban development in Ethiopia: challenges and policy responses. IUP J Governance Pub Policy 7(1):59–75
MUDHC (Ministry of Urban Development, Housing and Construction) (2007) Participation manual for urban planning. Addis Ababa—Ethiopia
MUDHC (2012) Revised standards for structure plan preparation and implementation. Addis Ababa—Ethiopia
MUDHC (Ministry of Urban Development, Housing and Construction) (2014) Urban plan preparation and implementation strategy. Addis Ababa—Ethiopia
MUDHC (Ministry of Urban Development, Housing and Construction) (2017) Street design standards for urban Ethiopia. Addis Ababa—Ethiopia
Murray MJ, Greer D, Houston SM, Murtagh B (2009) Bridging top down and bottom up: modelling community preferences for a dispersed rural settlement pattern. EUR Plan Stud 17:441–462. https://doi.org/10.1080/09654310802618101
MWUD (Ministry of Works and Urban Development) (2006) Structure plan manual. Federal Urban Planning Institute
Narayan D, Chambers R, Shah MK, Petesch P (2000) Voices of the poor: crying out for change. Oxford University Press, New York
Oktay D (2002) The quest for urban identity in the changing context of the city: Northern Cyprus. Cities 19(4):261–271. https://doi.org/10.1016/S0264-2751(02)00023-9

Olujimi J (2009) Planning standards as effective tools for development control. https://doi.org/10.13140/2.1.3835.6165
Omer K (2017) The rise and fall of international planning systems; the dilemma of flexibility and certainty. https://www.researchgate.net/publication/317590936. Accessed 20 Nov 2020
Oxley M, Brown T, Nadin V, Qu L, Tummers L (2009) Review of European planning systems, centre for comparative housing research. Leicester Business School De Montfort University. https://dora.dmu.ac.uk/bitstream/handle/2086/7536/NHPAU%20Planning.pdf?sequence=1
Parasuraman AP, Zeithaml V, Berry L (1985) A conceptual model of service quality and its implications for future research. J Mark 49:41–50
Pauchard A, Aguayo M, Peña E, Urrutia R (2006) Multiple effects of urbanization on the biodiversity of developing countries: the case of a fast-growing metropolitan area (Concepción, Chile). Biol Conserv. https://doi.org/10.1016/j.biocon.2005.05.015
PWCNT (Parks and Wildlife Commission of the Northern Territory) (2002) Public participation in protected area management best practice, parks and wildlife commission of the Northern Territory. https://environment.gov.au/system/files/resources/45209a20-1ab1-40d1-aa26-cf4838948782/files/public-participation.pdf. Accessed 29 July 2020
Rodela R, Udovč A (2008) Participation in nature protection: does it benefit the local community? A Triglav National Park case study. Int J Biodivers Sci Ecosyst Serv Manage 4:209–218. https://doi.org/10.3843/Biodiv.4.4:4
Sachdev S, Verma H (2004) Relative importance of service quality dimensions: a multisectoral study. J Serv Res 4(1):93–116
Shiferaw D (1998) Self-initiated transformations of public-provided dwellings in Addis Ababa, Ethiopia. Cities 15(6):437–448
Taylor N (1998) Urban planning theory since 1945. London. Thousand Oaks. New Delhi: SAGE Publications
Terfa BK, Chen N, Liu D, Zhang X, Niyogi D (2019) Urban expansion in Ethiopia from 1987 to 2017: characteristics, spatial patterns, and driving forces. Sustainability (Switzerland). https://doi.org/10.3390/su11102973
UN-Habitat (1999) Reassessment of urban planning and development regulations in African cities. http://digitallibrary.un.org/record/625038
World Bank (1996) The World Bank participation sourcebook. Washington, DC
Xie Z, Hall J, McCarthy IP, Skitmore M, Shen L (2015) Standardization efforts: the relationship between knowledge dimensions, search processes and innovation outcomes. Technovation 48. https://doi.org/10.1016/j.technovation.2015.12.002

Behailu Melesse Digafe is currently a lecturer and a Ph.D. student of Land Policy and Governance at the Institute of Land Administration at Bahir Dar University in Ethiopia. He holds a B.Sc. in Urban Planning and an M.Sc. in Urban Design and Development. He has worked for the Amhara Urban Planning Institute for over seven years and has participated in more than 30 urban planning and design projects as an expert, team leader and supervisor. Currently, he teaches undergraduate and graduate courses related to his fields of study. Behailu also served as the Director of the Architecture Department and was involved in several community service projects.

Dr. Achamyeleh Gashu Adam is Associate Professor of Land Governance at the Institute of Land Administration at Bahir Dar University. His educational background and research interests cover a broad range of related areas such as land governance, land policy, land administration, land tenure and use management, land development management, and land registration and information systems. Achamyeleh has served as Dean of the Institute of Land Administration at Bahir Dar University, and in various other positions within the university and outside of it. To date, he has published more than 20 scientific articles in peer-reviewed journals and more articles are forthcoming.

Dr. Gebeyehu Belay Shibeshi is an assistant professor and Vice Dean of the Land Administration Institute of Bahir Dar University in Ethiopia. He has specialised in cadastre and registration systems. He has a bachelors' degree from Alemaya University, an M.Sc. from KTH Royal Institute of Technology in Sweden, and a Ph.D. from the University of Natural Resources and Life Sciences (BOKU) in Vienna, Austria. Gebeyehu has also had 15 years of experience in the industry and served in different capacities in the Regional Bureau of Rural Land Administration and Use of the Amhara region in Ethiopia. He is now fully engaged in teaching, research, and community service activities at Bahir Dar University.

Dr. Mengiste Abate Meshesha is an assistant professor at Bahir Dar University. He has a B.Sc. degree in Hydraulic Engineering from Arba Minch University in Ethiopia, an M.Sc. degree in Water Science and Engineering from UNESCO-IHE Institute for Water Education in the Netherlands and a Ph.D. in Integrated Water Resources Management from Bahir Dar University in Ethiopia. He has also worked for the Bureaus of Public Works and Urban Development and Water Resources Development in the Amhara National Regional State of Ethiopia. He has served as a board member, committee member and consultant for various organisations and projects. Recently, Mengiste coordinated and provided professional input in the preparation of the Structure Plan for the city of Bahir Dar.

Chapter 10
Complementing or Conflicting Rationalities? How Self-Production Practices in Collective Spaces Can Shape Urban Planning: Insights from Maputo City

Milousa António

Abstract Spatial planning and governance in African cities are often framed and conceived through the formal-informal binary. This view has been responsible for negative connotations that increase urban populations' vulnerability. Moreover, it has been heavily criticised as presenting a reductive view of urban development. Alternative framings such as "alternative informality" and "self-production" have recently contested such views by proposing process-oriented approaches that recognise the legitimacy of informal praxis. However, research on self-production practices has tended to focus on the household or municipal level, neglecting what can be termed "collective space". This chapter explores the production and use practices within collective spaces based on research conducted in two peripheral neighbourhoods in Maputo in 2019. The findings highlight the role and legitimacy of self-production practices in collective space to provide services, consolidate local governance and substantiate urban development. It finds that the role of local residents and authorities in urban planning has only tentatively been accepted by official municipal-level planning agencies. The chapter will reflect on how collective space can better overcome local challenges beyond the household level and represent potentialities for inclusive and democratic planning. However, there are still many challenges in collective space that remain poorly addressed.

Keywords Self-production · Collective space · Governance · Inclusive planning

10.1 Introduction

As cities in the Global South have evolved into diverse urban landscapes, *informality* has become a fruitful field for urban theorisation (Acuto et al. 2019: 476). Urban informality has its most significant expression in the Global South. For example, in Sub-Saharan Africa (SSA), an average of 53.6% of the urban population lived in the

M. António (✉)
New University of Lisbon, Lisbon, Portugal
e-mail: milousa.antonio@edu.ulisboa.pt

G. Alem Gebregiorgis et al. (eds.), *Planning Cities in Africa*, The Urban Book Series,
https://doi.org/10.1007/978-3-031-06550-7_10

so-called *slums* or *informal settlements* in 2018 (World Bank 2021). This condition is predicted to increase with current population and urbanisation projections. SSA is expected to experience population growth of over a billion people by 2050, with the urban population accounting for 48 and 59% in 2030 and 2050, respectively (UNDESA[1] 2018, 2019). With most of the population expected to be urban by 2050, planning practices are at the centre of debates concerning urban resilience and sustainability.

Conventionally, however, debates regarding planning practices are often framed in dichotomous and segmented ways. This is partly due to entrenched institutional cultures, but more importantly, because of the long-conceived formal-informal urban divide in spatial development (McFarlane and Waibel 2016; Acuto et al. 2019). This dual conception of space is more explicit in post-colonial cities, such as Maputo, where the "officially-planned" colonial centre is considered formal and the periphery "informal".

Additionally, most debates on *informality*, whether "structuralist" or "legalist" approaches (Acuto et al. 2019: 477), gravitate thematically towards housing and economy. For a long time, planning practices have concentrated on either the household or broader municipal level, avoiding what can be termed collective spaces. Collective space (CS) was first proposed by Sola-Morales as spaces "not strictly public or private, but both simultaneously. These are public spaces that are used for private activities, or private spaces for collective use" (Sola-Morales 1992 quoted in Gheysen et al. 2019: 118). Researchers have argued that informality is a crucial lens to understand CS (Sadiq et al. 2020) as informal practices are often collective practices.

However, little work has been done to document CS's in cities of the Global South. Research on CS's has typically focused on Global North contexts, focusing on spaces such as car parks and shopping centres (Gheysen 2018; Gheysen et al. 2019). In Global South cities, CS arguably has distinct attributes.

Efforts to improve post-colonial urban peripheries, such as slum-upgrading programmes, have often focused on housing and infrastructure interventions, with recent integrated approaches systematically failing to address spaces of collective infrastructure use and production, ecosystem services and social networks in collective spaces.

This article explores recent debates on *urban informality* and *collective space,* to challenge the *formal-informal* approach to urban planning in post-colonial cities such as Maputo. Specifically, it aims to contribute to a more nuanced interpretation of these realities through a systematic and holistic understanding beyond binaries. The study focuses on the production of collective spaces in two neighbourhoods in Maputo. The discussion is developed in three main parts. Firstly: a conceptual framework on how *informality* as a concept has been challenged, and the emerging alternative framings attempting to transcend the binary and emphasising the importance of CS are discussed. Secondly: a background of Maputo planning and governance as a support for the understanding of practices. Finally, an examination of practices producing CS

[1] United Nations Department of Economic and Social Affairs.

centred on self-production of housing, service provision and co-produced initiatives and the challenges posed by governance issues and the spatial result, are presented in four empirical sections.

10.1.1 From Informality to New Approaches Transcending the Formal-Informal Binary?

In the last two decades, notions of "informality" are being questioned and challenged. The *formal-informal* binary, presumed as distinct and opposed realities, has proved fluid and continuous. While many institutions and practitioners look at the "informal" as an exception and the "other" of the formal (Acuto et al. 2019: 476), many scholars have shown that the situation is more heterogeneous. Indeed, many post-colonial cities with a strong history of spatial segregation (Andersen et al. 2015a, b; Melo 2016), have grown into diverse realities characterised by overlapping planning practices, uneven urban morphologies and heterogeneity in infrastructure and service provision.

Despite this, many post-colonial countries and cities inherited "formal" planning regulations and principles, which, in many cases, have continued almost unchanged or with minimal development through time (Watson 2009). Additionally, *slums* and *informal settlements*, as lexicons used to describe human settlements in the Global South, have proven to emphasise negative connotations, solidifying preconceived perceptions and overlooking potentialities and inherent capacities. Critiques of such views have given space to alternative framings, aiming to contribute to more open and context-specific, southern-based urban theory. This section presents new conceptualisations challenging the formal-informal binary. It leads to a debate on CS as an interface of multiple urban systems, where the formal-informal continuum is apparent.

10.1.1.1 Urban Informality Questioned

Urban informality is an ambiguous notion. Its definition is diverse in "specific epistemological frameworks, disciplines, time and place" (Banks et al. 2020: 225). For this discussion, I adopt the term as defined by Roy (2005) as a "system of norms that govern the process of urban transformation" (Roy 2005: 148). Urban informality concerning land use and urban services has been central in the ongoing shift to the Southern perspectives in planning theory. Here, many academics have critically questioned the binary view through several propositions.

For instance, Roy (2005: 148) argues that formal and informal are not isolated sectors but a series of transactions that connect different economies and spaces.

Additionally, Huchzermeyer (2011) elaborates on how globally legitimated political discourses surrounding informality can be subversive and increase the vulnerability of urban populations. She criticises how campaigns like "Cities Without Slums", based on a "misunderstood target to achieve cities free of slums" (Huchzermeyer 2011: 10) impact interventions by governments and international agencies. Others argue that informality is a politically based "act", to treat certain territories as an "exception" (Pratt 2019). Such perceptions imply notions of what is desirable, superior and globally accepted (Acuto et al. 2019: 476).

The second proposition is that in contexts of rapid urbanisation where the state lacks capacity, informal practices are, or should be considered as, "legitimate". Jenkins (2008), for example, by exploring the concepts of legality and legitimacy concerning urban development in Maputo, argues that "illegal" and "informal" practices are, in reality, more socially legitimate than "formal" practices of the state and market. Moreover, such practices happen due to a noticeable discrepancy between state interests and capacities and those of the private sector and civil society. According to Jenkins, such discrepancy results from:

> A weak state attempting to control growing market forces unsuccessfully, and with relatively long term legacies of ignoring urban development interests of the majority of urban dwellers, who resolved their situation through "informal" solutions, with links to the private sector, albeit exploitative. (Jenkins 2008: 178)

Furthermore, Ahlers et al. (2014) contend that discourses surrounding informality "say more about the authority to legitimate certain practices than describe the condition of that particular practice" (Ahlers et al. 2014: 2). She relates legitimacy with public authority, suggesting fluidity in how practices can shift from formal or informal through legitimisation by authorities, sometimes to "safeguard" interests (Ahlers et al. 2014: 7).

Both perspectives are closely related. They explore informality beyond generalisations, acknowledging urban diversity. Moreover, a central reflection from such discourses is imperative to move beyond a material-focused approach that has become a spatial metaphor and ignores other urban processes co- or collectively-produced or ambiguously situated.

10.1.1.2 Alternative Framings of Informality and the Relation to Collective Space

Following recent debates on informality, academics have put forward alternative framings of urban morphology and planning in cities of the Global South. The first is what Andersen, Jenkins and Nielsen refer to as *alternative formality* (Andersen et al. 2015b: 424). *Alternative formality* emerges from a reality where "informal" practices have, in fact, high social legitimacy. Such practices are legitimate because they are, in part, "collective forms of sociocultural organisation" which "orientate and guide land-use practices in many 'unplanned' areas" (Andersen et al. 2015a: 329). According to Andersen, Jenkins and Nielsen [2015b], such practices relate to the state and are dominant in spatial transformation in cities such as Maputo.

The second conception is what Raposo (2012) refers to as *self-production*, a process-oriented approach view that primarily sees urban transformations as based on incremental interventions by communities. This concept is rooted in Lefebvre's (1991) ideas surrounding the production of space, where space is socially produced, and involves three interrelated aspects: "representations of space", "spatial practices" and "spaces of representation" (Lefebvre 1991: 38). This approach centres on understanding territories through how they are transformed, considering both individual and collective contributions (i.e., both tangible interventions such as collective infrastructure and how such spaces are represented, perceived and lived). By using "self-produced" as an alternative lexicon, Raposo intends to build on the capacities of local communities while acknowledging the existence of material and functional challenges such as the lack of basic infrastructure and services. Underlying this is the ambition to emphasise the positive aspects as opposed to a pejorative perspective. According to Melo (2016), self-production is the major way in which the peripheral areas of Maputo have developed through processes typically established by the inhabitants, such as land sub-division and occupation.

Hence, self-production relates to processes where residents build their houses, and often infrastructure, largely outside of "formal" regulations. Here, a certain level of self and collective planning shapes the space (broadly conceived as "public") beyond the household. On the one hand, these territories are products of precarity, ineffective planning, legislation and policies and weak institutional capacity. On the other hand, it shows the resilience and imagination of poor communities to organise themselves. While inhabitant's autonomy and capacity to build and transform their habitat may be limited, their agency transcends the household scale to the "public" space where there are multiple infrastructures and everyday interactions occur—the CS. This dimension is relevant for a systematic and holistic view of urban areas.

Therefore, these new perspectives on informality emphasise the relevance of *collective spaces* as critical to understanding peripheral territories' production in the Global South. Here, private interventions are often interventions in "public" spaces, which subsequently have a collective value. It is this collective value that often secures the legitimacy of such urban morphology.

10.1.2 Collective Space: The Place of Conflicting and Complementing Rationalities

Collective space (CS) is said to play an "essential role in the development of urban areas" (Sadiq et al. 2020: 267). Conceptually, CS has been discussed largely in architectural studies when questioning the relationship between public and private spaces (Avermaete et al. 2006; Belingardi 2012; Gheysen et al. 2019). The term denotes an ambivalence and "lack of vocabulary" (Gheysen et al. 2019: 111) to describe public spaces that are re-appropriated by citizens and/or "private" spaces with inherently collective use-value. The authors addressing CS have been concerned about how such spaces are "represented" by practitioners and governance actors, and secondly how to describe the use and value of such spaces.

However, such analyses have largely been based in Western contexts where collective spaces have been considered in relation to pedestrian areas, car parks, shopping centres, etc., typically in "urbanised" areas. Little research has considered CS in cities of the Global South and particularly in "semi urbanised" peripheries. In Western contexts, CS's are typically seen as already provided with infrastructure, with debates focused on collective reappropriation. However, in cities of the Global South, collective spaces are often *produced* in the absence of state infrastructure provision, between the formal and informal interface. Thus, CS's are a crucial means for residents to access infrastructures and services. Arguably, collective production occurs in far greater diversity in cities of the Global South, such as Maputo, than observed by authors such as Prendergast (2013).

Building on alternative conceptualisations of informality, CS in contexts like Maputo is where individual and "collective" self-production practices encounter the practices of actors such as infrastructure/service providers. Furthermore, CS can be considered an interface of "conflicting rationalities" (Watson 2003). These rationalities, mainly of the state and population (Melo 2016), interplay or coexist, sometimes complementing one another, creating an ambiguity in the perception of responsibilities in CS's. CS is therefore understood in this article as spaces of collective use/interactions *and* production. This relates to collective infrastructure, mobility and housing and social goods such as leisure, education, culture and commercial activities. The approach to CS focused on self-production practices can move beyond the binary view and increase the inclusivity of urban planning processes.

In some sense, "collective space" speaks to the fragmented understanding of the urban landscapes and allocation of governance responsibilities. CS traverses the boundaries of private household practices and collective social services/infrastructures and their interactions. In practice, CS is (i) a space of shared (often poorly outlined and overlapping) responsibilities, (ii) heterogeneous and collectively managed infrastructure and (iii) a broader site of socio-ecological reproduction.

This chapter, therefore, aims to examine self-production practices, as an important mechanism of CS production in Maputo's peripheral neighbourhoods. By framing the analysis from a process-oriented perspective, the chapter seeks a more impartial and constructive approach to these territories. The chapter focuses on three aspects of CS in particular: (a) self-production of housing and its relation to CS, (b) collective service/infrastructure provision and (c) wider urban improvement interventions. The central question is: what can we learn from the production and use of CS in shaping urban planning in cities of the Global South? This discussion is presented in four empirical sections preceded by a background section: (1) perceptions about CS by different actors, (2) self-production practices concerning public services, (3) the local (municipal and neighbourhood) governance system and their potential role in articulating all practices and finally (4) a qualitative assessment to CS to identify challenges through materialities.

10.1.3 Methodology

This study is based on primary research conducted from January to March 2019 in two peripheral neighbourhoods in Maputo city, Bagamoyo and George Dimitrov. The research took a qualitative approach, including interviews of households and key informants, transect walks and participant observation. This approach was taken to (a) observe and document spatial characteristics, (b) identify spaces of collective production/use, (c) to elicit (different) perceptions concerning CS and (d) to document and understand the responsibilities and self-production practices of various actors in CS's. A total of thirty-three interviews were conducted: twenty with residents (ten in each neighbourhood) and thirteen with key informants from relevant urban service providers (such as water, electricity, solid waste management utilities/agencies) and also local management and neighbourhood leadership (including neighbourhood secretaries, senior members of the community, etc.). Additionally, transect walks were conducted in each neighbourhood with local leaders and informed inhabitants.

The choice of these neighbourhoods as a case study was based on three main criteria: a visible spatial diversity in spatial configurations (i.e., infrastructure, housing development), that they have existed through most phases of the city's spatial development, and that they continue to experience significant transformations. Multiple cases were chosen to compare local governance arrangements relating to collective spaces. Thus, the research is intended to build up from/build on different actors' perceptions of self-production in collective space.

10.2 Background

10.2.1 Urban Development in Maputo City: Spatial Transformations

Maputo city, the capital of Mozambique, is integrated into a metropolitan region which houses over three million people, nearly 30% of the national urban population (INE[2] 2017, 2019). While the spatial extent of self-produced areas has, so far, not being fully mapped, it is estimated that, until 2008, 87% of the residential territory in Maputo was self-produced, "partially developed and semi-urbanized" (Henriques 2008: 70). This area corresponds to the municipal periphery, where about 89% of the population lives (INE 2017). Here, the expansion of self-produced neighbourhoods has occurred over decades of profound socio-political and economic changes (Oppenheimer and Raposo 2007; Melo 2013).

Since independence, Maputo has seen a rapid expansion and consolidation of self-produced areas. Such growth has replaced or encroached upon previously semi-rural agricultural land, green and ecological areas, including sensitive ecosystems, such as

[2] National Institute of Statistics (*Instituto Nacional de Estatística*).

the mangrove on the east coast (Beja da Costa and Faria Ribeiro 2017; Beja da Costa and Jorge 2019). During the colonial period, the duality between the *cidade cimento-caniço* (cement-reed city) was established by the Portuguese colonial government, which intended to consolidate their control partly through the reinforcement of a highly unequal planning system (Melo 2013; Andersen et al. 2015a). According to Melo, at the end of the colonial period, the peri-urban area experienced a considerable expansion. As shown in Fig. 10.1, between independence (1975) and the end of the civil war (1992), land occupation in the so-called *caniço* city, experienced an enormous expansion. The following decade was marked by the consolidation of these new residential areas until the 2000 floods led to a rapid occupation of the northern territory.

Two main factors influence more recent spatial transformations. Firstly, large-scale infrastructure projects, such as the "circular" road on the outskirts of Maputo and the Maputo-KaTembe Bridge, have encouraged land occupation in surrounding areas with high ecological value and environmental risk. Secondly, the more incremental and continuous *home-making* processes across self-produced neighbourhoods, as examined by Jenkins (2012) and Andersen (2013) in the *homespace* project.

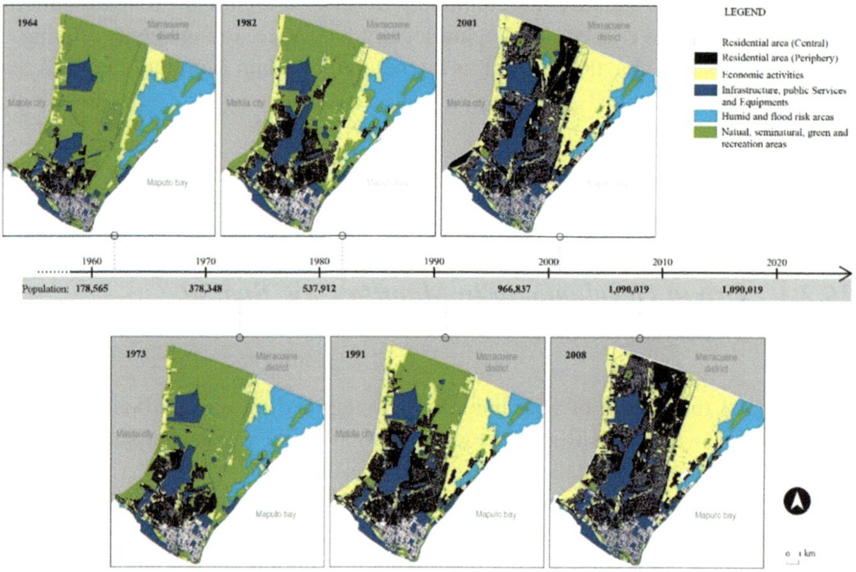

Fig. 10.1 Land occupation and the expansion of residential areas in Maputo between the 1960s and 2000s (*Source* Image edited by the author, from the land use maps by Henriques [2008: 127])

10.2.2 Planning and Local Governance

Urban development in Maputo has been influenced only in a limited sense by official planning documents and practices. This is exemplified by a history of failed attempts to implement land-use planning, limited by "human resources, institutional capacities and lack of political will" (Andersen et al. 2015a: 339). In practice, complex forms of overlapping planning practices (Andersen et al. 2015a), along with a normalised illegal land market (Perriard 2017), dominate land-use development. This reality creates conditions for unofficial planning practices, sometimes replicating official patterns, termed "inverse governability" by Nielsen (2011: 332). The current municipal planning system remains a legacy from the colonial period characterised as rigid, expensive, lacking reference to sustainable development and detached from local contexts of urban change. While this conundrum in Maputo seemingly discredits the value of "formal" planning, planning systems have been said to work in the interests of groups such as the ascending middle class (Mazzolini 2016) and others that see the opportunity for forms of real estate speculation.

In 1996, Mozambique adopted a decentralised process to achieve "good governance" and increase service delivery, a strategy heavily promoted by international organisations and donors. However, implementation has been uneven, with many services remaining centralised (Silva 2016). Structurally, the municipal authority is constituted of the "urban", "district" and "neighbourhood" levels, as shown in Fig. 10.2. While neighbourhood authorities are the primary interface with residents and have mobilisation power, they play an insignificant role in urban planning. Crucially, debates on decentralisation have often disregarded or failed to address the emergence of bottom-up governance arrangements emerging through collective agency and the blurred lines between broader governance actions/institutions and alternative practices.

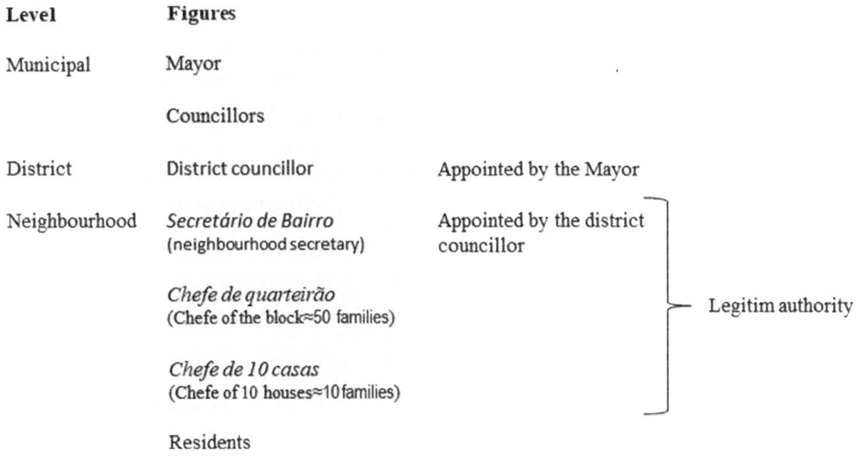

Fig. 10.2 Local administrative structure (*Source* Author)

Nevertheless, the city has allowed research and experimentation from various actors, increasingly recognising the inhabitant's participation; many of them attempting context-based, bottom-up solutions for housing, infrastructure and commerce. Interestingly, in recent upgrading interventions, authorities are experimenting with participatory processes of recovering public land for infrastructure development, for instance, in the Chamanculo C requalification project [interview 13, 30].

10.2.3 Case Study: Bagamoyo and George Dimitrov

The case study is of two adjacent neighbourhoods, Bagamoyo and George Dimitrov, located on the periphery of Maputo city along one of the main national arterial roads. They are also both in the Infulene basin, an important area or urban agriculture. The areas existed before independence and were jointly known informally as "Benfica". According to an individual who has lived in the area since 1950, in the late 1960s, a Portuguese investor named Custódio da Graça developed infrastructures, housing and services close to what is today known as George Dimitrov. Later, colonial buildings to house some public institutions were built in the area. However, the most significant development in land occupation took place during the civil war between 1977 and 1992, when refugees were looking for safer areas to live while the conflict prevailed. They established themselves in the area with intermediation, or as indicated by one community leader, "permission from the *Regulo* and *Secretários de Bairro*" [interview 33].[3] The first is a traditional authority, and the second is a community authority established by FRELIMO,[4] later legitimised as an intermediate figure between the state and citizens (Meneses 2009).

Spatially, the neighbourhoods are interesting territories to study CS due to their morphological and functional diversity, particularly concerning the mobility system and areas with high (although neglected) ecological/economic importance. Figure 10.3 shows the location of both neighbourhoods. In Maputo, the spatial dimension of self-produced territories is distinguished (by the municipality) in two categories: "planned" and "unplanned". "Planned" settlements represent areas with an ordered (although not regularised) morphological structure, whereas "unplanned" areas are spontaneous and more organically structured (CMM[5] 2016: 23). Here, the planned areas are related to unofficial planning practices described by Nielsen (2011).

---

[3] Interviews have been numbered and names withdrawn to protect the anonymity of interviewees.

[4] Mozambique Liberation Front (*Frente de Libertação de Moçambique*), the political party governing Mozambique since independence in 1975.

[5] Maputo Municipal Council (*Conselho Municipal de Maputo*).

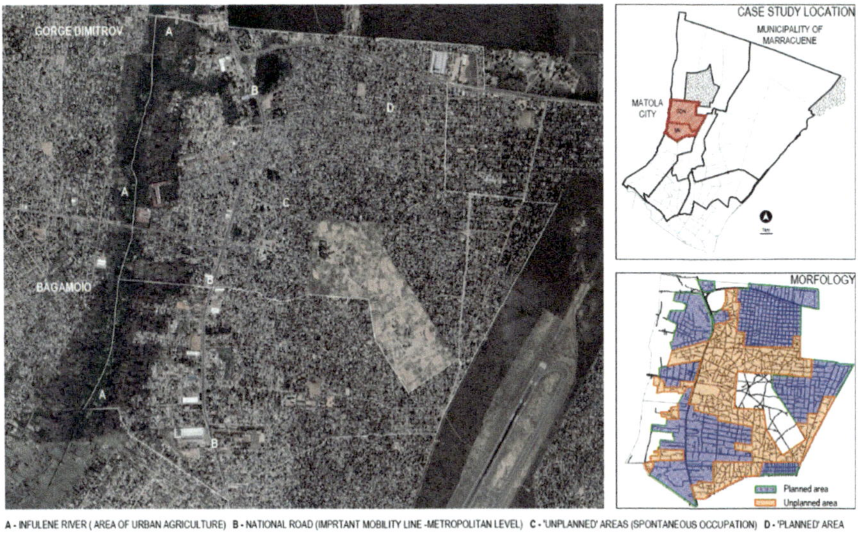

Fig. 10.3 Case study location and some important urban pre-existences (*Source* Adapted by the author from António 2019—images on the right; Author, based on aerial photographs taken in Google Earth—image on the left)

10.3 The Production of Collective Space in Maputo

Collective space (CS) in the context of Maputo is broadly produced through three related processes. Firstly, the self-construction of housing, a process through which individuals (inhabitants), through cooperation with local neighbourhood authorities, shape and reshape public spaces' broader spatial/material condition. Secondly, through service/infrastructure provision, whether by public or private service providers, who consolidate such spaces through characteristically heterogeneous infrastructures. Thirdly, through co-produced improvement actions through which various actors (including central and municipal governments, international agencies, and residents) experiment with urban solutions.

In the following section, I explore critical aspects of the production and use of CS, including how inhabitants, planners and service providers perceive these spaces, the social practices and processes shaping CS, and how such processes are governed.

10.3.1 Conceiving Collective Space: Perceptions and Responsibilities

Recognising the perceptions of actors producing and using CS is a crucial way to understand their agency. As a starting point, I examine perceptions surrounding the self-production of housing. Self-production practices in housing are relevant to perceptions of CS because housing development in Maputo is largely an incremental, unlicensed process driven by inhabitants. Residents are responsible for constructing houses, initiating the administrative procedures for accessing land (ultimately owned by the state), and establishing the urban form and functional patterns. These processes are decisive in producing CS. While, in principle, housing development should follow "ordered" rules, there is, in practice, great diversity in how houses are built and, therefore, how space is configured, as can be seen in Fig. 10.4. This is partly a result of how residents perceive CS. Such perceptions critically influence housing construction decisions in confrontation with CS.

Inhabitant's perceptions of CS are evident in the initial arrangements of land occupation. A common theme among the household interviews was the land occupation processes concerning the space beyond households. Here, the agency and the intermediary role of local leaders and informal land market agents also influence occupation processes. For instance, most residents from "planned" areas stated that they (and their neighbours) respected CS because "when occupied, the area was already parcelled out, thus everyone built their construction/fence within the established limits" [interview 10]. Furthermore, some understand that "the land subdivision was undertook by the municipality" [interview 11], others acknowledged that when they "exceed the limit, the municipality might come, sometimes to the point of demolition" [interview 9]. Thus, although there is a broad understanding that "the authorities" control spaces beyond the boundaries of houses, most factor CS into house building decisions, and as one respondent indicated, such space "belongs to their backyard" [interview 11, 12]. So, if "a person intends to use it [the area in front of the house], they must ask the owner [the house owner nearest to the collective space] for permission" [interview 13].

B - Bagamoyo neighbourhood, **GD** - George Dimitrov neighbourhood. **(a)** 'planned' area, **(b)** 'unplanned' area.

Fig. 10.4 Distinct public space morphologies resulting from self-production of housing (*Source* Author)

Concerning residents in "unplanned" areas, there was a broad agreement among many interviewees that official boundaries were not respected: "not everyone considers public space and space for other infrastructure, some even wrap the electricity pole inside the backyards" [interview 13]. Respondents typically equated such phenomena to the land occupation process:

> My backyard will be partially demolished in the future rehabilitation of the road. It is the fault of the person who (re)sold the space who was not careful and indicated an area within the road. [interview 13]

Referring to the same case, a separate interviewee described, "perhaps there was a lack of information from those superior to me [local neighbourhood leaders] why today some houses are destroyed to make room for the road" [interview 2].

Such perceptions highlight a crucial dynamic in how the governance and land occupation process influences inhabitants' perceptions of CS. Uncertainty over the limits between private and public space can lead to "conflicts over plot limits" [interview 15, 16]. Moreover, residents actively make "interventions" in such spaces. As one interviewee described, when "everyone decides to put obstacles on the street" [interview 6], it changes the spatial dynamic of public spaces over time. Such "obstacles" are typically built features, mostly for security or accessibility. Residents tend to portray the area beyond houses as an infrastructure and "car-oriented" place, meaning that considerations must be made to leave just enough space for cars and other infrastructures when building houses. Thus, although there are in principle definite limits to demarcate private houses and public spaces, in practice, there is some degree of negotiation and ambiguity in how these spaces are produced.

When asked about the importance of CS concerning mobility and leisure, a common theme across interviews was that such spaces were "very important" for "health" [interview 10], the "development of certain physical and mental abilities" [interview 8], and because they "serve many purposes, be it meetings, or any kind of socialising" [interview 2], and especially "for children to have fun" [interview 14]. One respondent stressed that collective spaces were important for "young people who are in a bad life" [interview 3]. Respondents emphasised, however, certain risks in CS's such as "being overrun by cars" [interview 11], because, although they valued such spaces, they were managed poorly. Interestingly, there is an understanding that historically, the authorities failed to preserve relevant areas for CS. As one respondent stated, "the municipality only preserved public land for football fields" [interview 4].

Municipal level planning authorities and service providers had different perceptions of collective space, more grounded in the formal-informal dichotomy. Overall, there is tacit acceptance or toleration of "unlicensed" activities in collective spaces among most service providers and planning authorities, except for the municipal commerce department that has a more regulatory, anti-informality position. I also found evidence of an "eradication" discourse common across planning bodies that see practices outside their regulation as needing to be converted/eliminated [interview 21, 23, 25]. Interestingly, many interviewees stated that the formal authorities are responsible but largely absent in practice when asked who is responsible for self-produced spaces. Key informants acknowledged how actors such as residents, land market mediators and neighbourhood leaders had, in practice, taken on many responsibilities for some "planning" actions [interview 21]. This view differed for those responsible for service provision (water and electricity), who generally had a more tolerant and practical idea about self- and collective-production interventions to infrastructure. As such, while planning bodies should, in principle, coordinate practices in CS, a much more fragmented approach to spatial development and planning is evident.

10.3.1.1 Responsibilities and Collective Space

Residents understand responsibilities surrounding CS in as many different ways as they conceive of them. When asked about responsibilities for the provision and maintenance of public services, although many indicated that residents had a level of responsibility, it is widely understood that "this is the subject of local leaders" [interview 20]. Very few residents were able to indicate the corresponding actor responsible for different services, except for potable water and electricity, which were more clearly seen as the responsibility of FIPAG's,[6] and local (private) providers and EDM's,[7] respectively. In their uncertainty, many residents ascribed a general responsibility to local leaders.

Furthermore, neighbourhood leadership is central to how residents perceive their collective responsibilities and participate in or contribute to local development. As shown, there is a considerable density in the number of community representatives. If we consider *chefe de quarteirão* (block chief) and *chefe das dez casas* (chief of ten houses), there is approximately one leader for every fifty residents. Neighbourhood leaders act as mediators for residents on formal procedures and for gaining access to services, and represent the community's voice in local development matters.

However, according to some local leaders, there are challenges concerning land occupation because residents, when overreaching their temporary limits to build permanent fences, frequently and knowingly take common land [interview 32 and 33]. According to one leader, many residents ignore directions and build quickly because they recognise the authorities' dilemma regarding the destruction of individual infrastructures. Concerning participation, all interviewed leaders acknowledged that residents are widely involved in local decision-making processes. However, residents, especially the youth, often feel excluded. For example, they feel that "meetings that are held in the neighbourhood are to inform or discuss specific problems, rather than ideas" [interview 18]. In summary, the perception of diverse responsibilities, with an emphasis on the role of local leaders, shapes local planning practices and contributes to understanding power dynamics in CS.

[6] Water Supply Investment and Heritage Fund (*Fundo de Investimento e Património do Abastecimento de Água*), a national public institution responsible for promoting water supply services.

[7] *Electricidade de Moçambique*, a state-owned utility company, responsible for electricity services.

10.3.2 *Diverse Self-Production Practices Concerning Public Services*

Collective space (CS) is a product of self-production processes and practices that respond to the need for various public services. One service is security. Wealthier and more resourceful residents were observed to be less open concerning collective space, in that they often built large walls and other security features that served as a barrier. While household security is the dominant underlying rationale, security in collective spaces themselves was also an important concern among respondents. Individual interventions are made, such as lighting within people's plots that also light surrounding areas, which impact collective safety.

Moreover, residents build infrastructure in CS to address challenges such as drainage (Fig. 10.5). Other infrastructure and built interventions include modifying roads to improve access to commercial establishments or providing better access for vehicles. Many of these interventions are collectively organised, financed at a community level and solve collectively experienced issues. Consequently, there is a dynamic interplay between accessibility and safety in common areas. The quality of such interventions reflects the profile of the residents involved. They are often temporary solutions. Neighbourhood authorities typically support such interventions, as they understand community needs and acknowledge the lack of resources at the municipal level to provide key services.

Services such as water, electricity and solid waste management, are provided by "official" agencies. They are relevant not only to individuals but also to collective spaces. Firstly, water provision was, until 2007, dominated by boreholes, private

(a) Intervention on a road infrastructure by residents in 25 de Junho A. a neighbourhood next to Bagamoyo. Images by Dercisio Tembe, 2019

Fig. 10.5 Individual and collective interventions on CS by residents (*Source* Modified by the author from António [2019: 80–93]) vendors and shared water taps (Macucule 2018). These provision methods are typically situated in collective space. Importantly, such water provision was technically "informal" until legitimised by the state in 2015, and today represents a recognised, sometimes favoured, option for residents [interviews 4, 11, 13, 16 and 33]. Existing decentralised water taps are increasingly recognised by water authorities as "an infrastructure that belongs to no one" [interview 25]. Recently, demand is mainly met by private providers whose local residents viewed to offer a more reliable and efficient service. Moreover, water infrastructures are managed by multiple actors.

Regarding electricity, street lighting is one of the most critical components in collective spaces. In both neighbourhoods, the distribution and availability of street lighting is uneven. Spatially, areas of "spontaneous occupation" are generally less well served than those of "ordered occupation". This is relevant to neighbourhood security, one of the most pressing challenges. When asked what aspects are in need of urgent improvement in the neighbourhood, half of the interviewed residents highlighted street lightning, some based on the security imperative [interviews 7, 8, 9, 10, 11 and 14], others concerned with quality [interviews 2, 13, 16 and 20]. Residents have long practised self-producing and self-maintaining electricity infrastructures themselves, including installing polls/wire, building/repairing lampposts and lamps. Others produced their own street lighting (Fig. 10.6).

Solid waste management is an iconic case of self-production practices that have been legitimised and recognised by many actors, including the municipality. In principle, the municipality manages solid waste. This includes two levels of services: the neighbourhood level, the primary collection (from houses and public spaces), and the secondary collection from the containers to the municipal dump. The first is operated by a local microenterprise chosen through public tender. The second is a larger company, which operates at the municipal level. Both are contracted by the municipality (Macucule 2018). Essentially, this waste collection system mimicked and incorporated aspects of self-production practices that already existed.

This service is considered innovative because of the financial sustainability that has been achieved [interview 22]. Taxes are collected through electricity bills to fund the service. In addition to the public service, which is viewed as satisfactory, residents, along with neighbourhood leaders, enforce the maintenance of the neighbourhood through "sporadic block cleaning days" [interview 33] (Fig. 10.7).

Fig. 10.6 Self-produced (potential) solutions for street lighting (*Source* Author)

Fig. 10.7 Microenterprise operating in a neighbourhood (*Source* Modified by the author from António [2019: 83])

It is clear that self-production of services is crucially important for collective space, where residents and local leaders play a key role. What is shown is a predisposition of residents to contribute to local development through providing and managing public services. The service provision dynamics described above show a fragmented but increasing proliferation of efforts by the public actors to capitalise on, mimic and legitimise creative self-produced practices in CS.

In the context of this reappropriation and legitimisation of practices by the authorities, co-production (mostly experimental) initiatives have been tested in these territories. For instance, the "George Dimitrov neighbourhood improvement project" (*Projeto de melhoramento do bairro George Dimitrov* 2015), an intervention that attempted to integrate informal commerce, inspired by *informal* practices. Also, an ongoing (since 2015) university-led initiative named *Kaya Clínica* offers technical assistance (as mediators for the municipality) to residents of George Dimitrov to gain access to DUATs[8] through regularisation, and therefore a potential platform to ensure the balance between individual and common land.

The above interventions represent attempts by the state to address issues in self-produced areas. However, such initiatives primarily depend on foreign funds and have an insignificant territorial impact, as they fail to be replicable.

10.3.3 Governance of Collective Space

As argued, there is a "dense" system of governance and control at the local level. This system has benefits and downsides that became evident in the research. Among the benefits, neighbourhood leaders are locally recognised figures. This gives residents a high degree of mobilisation power, which provides convenient conditions for engagements. For instance, in co-produced interventions where residents have to

[8] *Direito do Uso e Aproveitamento de Terra*, a title (document) that formalises the right to occupy land.

give up part of their plots for public space, the existent trust between these authorities and residents results in fewer conflicts.

However, although considered legitimate, local authorities are not formally bound to the municipal planning and management system, nor are they trained to assume spatial planning tasks. One of the constraints is their direct/indirect involvement in some unofficial land occupation processes, including the occupation of public reserves and ecological areas. Another obvious constraint is uneven management. The "expected mediation" role relies on the understanding of the leader representing residents, which largely impacts how residents participate in local development. When residents were asked about their involvement in improving living conditions, opinions split between those who did not believe they had a voice regarding existing space [interviews 4, 5, 8, 17, 20] and those who indicated that local authorities are those who represent them on such occasions [interviews 11, 13, 19].

One critical governance challenge relates to the perception of responsibilities. The state often uses the complexity of responsibilities to avoid their obligations. Another key challenge concerns power dynamics. Despite the decentralisation process, authorities at the neighbourhood level have little power in relation to some centralised services, adding complexity to the governance.

10.3.4 Assessing Challenges in Collective Space

To assess the practices in the production and use of collective space (CS), a transect walk was conducted across neighbourhoods with local leaders to identify major challenges in CS (Fig. 10.8). The following discusses the dynamics between different services in distinct geographical areas in the neighbourhoods, allowing for some critical observations. The first is the interdependency between different infrastructure systems within collective space, including those with high ecological value. The mismanagement or inefficiencies of one infrastructure can affect another. For example, I observed that the lack of adequate sanitation and drainage systems in one part of the neighbourhood directly impacts the water quality in another part. Moreover, there are critical social challenges in CS. For instance, uneven distribution creates social challenges such as security with regard to public lighting and other social goods. Additionally, for some infrastructural deficiencies, residents sometimes develop only short-term solutions.

These challenges suggest the need for better understanding of the problems to address challenges within CS. However, at the institutional level, common issues are addressed in a segmented way. Therefore, municipal authorities need to take a more flexible approach to self-production practices beyond individual household needs. In fact, there is an urgency for a more systematic approach that realises the potentialities of self-produced practices in CS so as to meet current needs and aspirations.

This is an initial assessment of the reality of collective spaces, and is based on the assumption that self-production practices play a significant role in the dynamics of urban life in Maputo. Here, I argue that self-production practices transverse the

	Analised system	Location	Geographical area: Ecological value	National Road	'Planned'	'Unplanned'
Infrastructure and Services	Water	BN / GDN	Water pollution, caused by domestic, commercial and industrial affluents (lack of drainage system)	Irregular access to potable water; Low Infrastructure mantainance elevating risk of water contamination	Some risk of contamination of drinking water by the eminent exposure of the infrastructure, partially caused by the predominance of unpaved streets	High risk of drinking water contamination by the (frequent) exposure of the infrastructure, partially because of unpaved streets, but also the street configuration and conflict with other infrastructures
	Drainage (and Sanitation)	BN / GDN	Insufficient and inadequate infrastructure increasing the risk of flooding	Discontinuity in infrastructure maintenance, causing frequent localised flooding	Shortage of drainage infrastructure causes occasional erosion situations with effects on mobility. Frequent individual and collective interventions by residents (near their homes) to face floods and allow better accessibility.	
	Solid waste	BN / GDN	Regular cleaning by small producers and cooperatives	Secondary collection irregularity, causing unhealthiness. Some areas of commerce coexist with polluted air	Primary (household) collection and cleaning of public spaces by the micro company. Also, cleaning routines by residents, on a community initiative basis.	
	street lightning	BN / GDN	Inefficiency of existing public lighting, compromising public space sefety	Continued and consistent service	Inefficiency, sometimes inexistency of public lighting infrastructure, compromising public space sefety	Deficiency in public lighting infrastructure, compromising public space sefety. Pontual interventions by residents
	collective equipments	BN	Lack of leisure and recreation equipment	Inadequate conditions for pedestrian circulation and insufficient infrastructure for passenger transport	Existence of health and education equipment. However, Insufficiency of leisure and recreation equipment and disarticulation with the transport system. Residents walk long (not safe and uncomfortable) distances to reach some equipment	
		GDN			Lack of health equipments, and insufficient leisure and recreation equipment and disarticulation with the transport system. Residents walk long (not safe and uncomfortable) distances to reach basic services	

BN Bagamoyo neighbourhood
GDN Gorge Dimitrov neighbourhood

Fig. 10.8 Results from transect walk in Bagamoyo and George Dimitrov neighbourhoods (*Source* Modifyed by the author from António [2019: 101])

household level in response to service needs, heavily shaping the importance and need for CS. It is important to stress that, through this case study, the ambiguity of the formal/informal binary takes shape through those who hold responsibilities and the way CS is perceived, lived and governed. Although its practices are highly legitimate, this does not negate the importance of the challenges faced.

10.4 Conclusion

In conclusion, alternative framings of *informality* are a useful starting point for a less pejorative understanding of urban peripheries in the Global South. They offer a lens through which to understand collective space. Lefebvre's (1991) triadic conception of space opens up potential frameworks that embrace the complexity of these spaces and their production practices, as underlined by the concept of self-production. I used *self-production* as a lens to identify, outline and assess practices that create and transform CS in Maputo, from which I made some critical reflections. While this is a valuable lens, we might also consider *collective*-production, which emphasises actions that are rarely self-referential but emerge because of, and have repercussions for CS.

Self- and collectively-produced practices are more extensive than previously understood. Such practices allow for the survival of inhabitants as they use their imagination to find solutions for everyday challenges such as access to services, safety, building more liveable neighbourhoods and navigating a complex landscape of political and planning "authority". Through self- and collective- production practices, residents address individual, "private" concerns as well as collective concerns. The case of Maputo suggests that such a situation arises partly because planning is not a political priority, and authorities are passive (or reactive) towards improving CS with limited capacity. Although there has been some recent legitimisation of self-produced service provision practices, there remains an ambiguity concerning responsibilities in CS's, which to a certain extent allows the state to ignore their role in this domain.

This study shows that CS is a product of alternative use *and* production practices. This may speak to an alternative understanding of CS in cities of the Global South, which contrasts with previous research focused on Western contexts (see Gheysen 2018). Practices such as local solutions to street lighting, provide clues for necessary regulatory adjustments to shape planning based on the realities "on the ground". Focusing on CS shifts debates on informality from the housing and economy scale, and indeed dichotomous understandings of the production of space, to the interface and interaction between different social, technical and governmental (complementing) processes. To further understand how such practices in CS can shape urban planning, further case studies are needed on collective spaces, particularly in Global South contexts.

Acknowledgements This research was funded by Instituto Camões (Camões—Institute of Cooperation and Language, IP). I thank Professors Margarida Pereira (new University of Lisbon-FCSH) and Professor Isabel Raposo (University of Lisbon-FA) for their support during the research. I also thank Dr Shaun Smith (Utrecht University) for his support in structuring this chapter. Finally, I thank the reviewers for their helpful feedback and comments.

References

Acuto M, Dinardi C, Marx C (2019) Transcending (in)formal urbanism. Urban Stud 56:475–487. https://doi.org/10.1177/0042098018810602

Ahlers R, Cleaver F, Rusca M, Schwartz K (2014) Informal space in the urban waterscape: disaggregation and co-production of water services. Water Altern 7:1–14

Andersen JE (2013) Urban morphology, urban transformations and conservation in Maputo, Mozambique. Lessons learnt and transfer of knowledge to other regions. Urban Development Symposium. Iraq

Andersen JE, Jenkins P, Nielsen M (2015a) Who plans the African city? A case study of Maputo: part 1—the structural context. Int Dev Plan Rev 37:329–350. https://doi.org/10.3828/idpr.2015.20
Andersen JE, Jenkins P, Nielsen M (2015b) Who plans the African city? A case study of Maputo: part 2—agency in action. Int Dev Plan Rev 37:423–443. https://doi.org/10.3828/idpr.2015.25
António M (2019) *Sustentabilidade em bairros autoproduzidos de baixa renda. Abordagem focada no espaço coletivo: O caso de George Dimitrov e Bagamoio em Maputo* [Sustainability in low-income self-produced neighbourhoods. An approach focused on collective spaces: The case of George Dimitrov and Bagamoio in Maputo] (p. 128). Dissertation (in Sustainable Urbanism and Spatial Planning)—School of Science and Technology and School of Social Sciences and Humanities. University of Lisbon, Lisbon.
Avermaete T, Hooimeijer F, Schrijver L (2006) Urban formation and collective spaces. OASE 71
Banks N, Lombard M, Mitlin D (2020) Urban informality as a site of critical analysis. J Dev Stud 56:223–238. https://doi.org/10.1080/00220388.2019.1577384
Beja da Costa A, Ribeiro LPF (2017) Mangroves of Maputo, Mozambique: from threatened to thriving? Plan J 2:629–651. https://doi.org/10.15274/tpj.2017.02.02.21
Beja da Costa A, Jorge S (2019) A landscape in mutation: the case of Bairro dos Pescadores in Maputo, Mozambique. Finisterra 54:145–162. https://doi.org/10.18055/finis17838
Belingardi C (2012) Collective and common spaces making city. In EURAU 12 Porto/Public Space and Contemporary City—6th European Symposium on Research in Architecture and Urban Design. Porto
CMM (Conselho Municipal de Maputo) (2016) Manual Metodológico de Intervenção em Assentamentos Informais. Maputo
Gheysen M (2018) Unlocking the potential of collective spaces in the peri-urban condition: the case of South-West Flanders. In: The horizontal metropolis between urbanism and urbanization. Springer International Publishing, pp 351–359. https://doi.org/10.1007/978-3-319-75975-3_34
Gheysen M, Van Daele E, Scheerlinck K (2019) Collective space in the peri-urban condition: opposing oppositions in all city/all land. In: Reicher C, Bayro-Kaiser F, Tawil M, Shawash J, Bäumer K, Polivka J (ed) Reclaiming public space through intercultural dialogue. LIT Verlag
Henriques C (2008) Maputo Cinco décadas de Mudança Territorial. O uso do solo observado por tecnologias de informação geografica. Instituto Português de Apoio ao Desenvolvimento, Lisbon
Huchzermeyer M (2011) Cities with slums: from informal settlement eradication to a right to the city in Africa. Juta Acade
INE (Instituto Nacional de Estatística) (2017) Recenseamente Geral da População e Habitação—IV RGPH 2017. Divulgação dos resultados preliminares. Maputo
INE (Instituto Nacional de Estatística) (2019) IV Recenseamento Geral Da População E Habitação 2017. Resultados definitivos. Maputo
Jenkins P (2008) Investigating the concepts of legality and legitimacy in sustainable urban development: a case study of land use planning in Maputo, Mozambique. Hum Settl Dev IV: 1–8
Jenkins P (2012) Home space—executive summary. Synthesis Report. Home Space Maputo. http://homespace.dk/publications.html. Accessed 10 June 2020
Lefebvre H (1991) The production of space. Basil Blackwell, Cambridge. https://doi.org/10.2307/490789
Macucule DA (2018) Provisão de serviços urbanos em Maputo
Mazzolini A (2016) The rising 'floating class' in Sub-Saharan Africa and its impact on local governance: insights from Mozambique. In: Silva CN (ed) Governing urban Africa, vol 45. Palgrave Macmillan, pp 213–246. https://doi.org/10.1057/978-1-349-95109-3
McFarlane C, Waibel M (2016) Urban informalities: reflections on the formal and informal. Routledge, New York
Melo V (2013) Urbanismo português na cidade de Maputo: passado, presente e futuro. URBE - Revista Brasileira de Gestão Urbana 5:71–88. https://doi.org/10.7213/urbe.7786

Melo V (2016) The production of urban peripheries for and by low-income populations at the turn of the millennium: Maputo, Luanda and Johannesburg. J South Afr Stud 42:619–641. https://doi.org/10.1080/03057070.2016.1196955

Meneses MPG (2009) Poderes, direitos e cidadania: O 'retorno' das autoridades tradicionais. Revista Crítica de Ciências Sociais 87: 9–41

Nielsen M (2011) Inverse governmentality: the paradoxical production of peri-urban planning in maputo, mozambique. Crit Anthropol 31:329–358. https://doi.org/10.1177/0308275X11420118

Oppenheimer J, Raposo I (2007) Subúrbios de Luanda e Maputo. Edições Co

Perriard ED (2017) Land tenure tensions in Maputo a study of the neighbourhood of Polana Caniço. L'Afrique en questions 38:1–7

Pratt A (2019) Formality as exception. Urban Stud 56:612–615. https://doi.org/10.1177/0042098018810600

Prendergast A (2013) Dinàmiques urbanes dels barris sub i periurbans de Maputo: reflexions sobre l'espai col·lectiu generat. QRU: Quaderns de Recerca en Urbanisme. https://doi.org/10.5821/qru.9545

Raposo I (2012) Bairros de génese ilegal: metamorfoses dos modelos de intervenção. In A Cidade entre Bairros. Caleidosco, pp 107–119

Roy A (2005) Toward an epistemology of planning. J Am Plann Assoc 71:147–158. https://doi.org/10.1080/01944360508976689

Sadiq A, Scheerlinck K, Pak B (2020) Studying the interrelationship of the formal and informal processes in the making of collective spaces: the case of Place Liedts and environs, Schaerbeek, Brussels. In: Sadiq A, Scheerlinck K, Pak B (ed) Communities, land and social innovation: land taking and land making in an urbanising world, pp 263–281. https://doi.org/10.4337/9781788973779.00022

Silva CN (2016) Governing urban Africa. Palgrave Macmillan. https://doi.org/10.1057/978-1-349-95109-3

UNDESA (United Nations Department of Economic and Social Affairs) (2018) World Urbanization Prospects 2018. Webpage

UNDESA (United Nations Department of Economic and Social Affairs) (2019) World population prospects 2019. Department of Economic and Social Affairs. World Population Prospects 2019

Watson V (2003) Conflicting rationalities: implications for planning theory and ethics. Plan Theory Pract 4:395–407. https://doi.org/10.1080/1464935032000146318

Watson V (2009) Seeing from the South: refocusing urban planning on the globe's central urban issues. Urban Stud 46:2259–2275. https://doi.org/10.1177/0042098009342598

World Bank (2021) Population living in slums (% of urban population)—Sub-Saharan Africa. Data. https://data.worldbank.org/indicator/EN.POP.SLUM.UR.ZS?locations=ZG. Accessed 2 Aug 2021

Milousa António is a Ph.D. student of urbanism at the University of Lisbon. She received her Master of Sustainable Urbanism and Spatial Planning from the New University of Lisbon. She has served as a researcher on numerous international projects, including projects relating to climate change initiatives, and has collaborated on the AFRICA HABITAT project led by Isabel Raposo. At the Centre of Studies for Development of Habitat (CEDH) in the Faculty of Architecture and Planning at Eduardo Mondlane University, she has worked as an assistant lecturer and researcher. Her research interests focus on spatial planning and sustainable urban development in cities of the Global South.

Chapter 11
Translating Globalised Ideals into Local Settings: The Actors and Complexities of Post-settlement Water Infrastructure Planning in Urban Ghana

Francis Dakyaga

Abstract Following the principles of the networked city and urban planning, proactive planning of water infrastructure is pertinent for attaining universal water access. Ironically, in cities of the Global South, water infrastructure provision takes the form of post-settlement networks—where human settlements evolve to steer the provision of the large-scale water network. However, little is known about the complexities, the processes and motives, the actors involved and how they navigate towards universalising water access. I investigate this kind of infrastructure planning ideal, drawing inspiration from technological translations from the Global North to the Global South, using the case of Wa, a secondary city of Ghana. The study revealed that off-grid water systems initially served water in secondary cities. The large-scale water network later evolved as a "reactive measure" driven by the rise in population, and the failure of the off-grid water infrastructure to attain universal water access. Despite that, resistance from residents, spatial disorder and sprawling growth, utility policies and in capabilities challenged the efforts of the state utility towards attaining a universal water supply. Through creativity, the utility providers negotiated and invented multiple models of water supply contradictory to the "mono-modal" principles of the networked city. This produced and segregated water access across the urban zones of the city. The findings suggest that though the post-settlement water network provision represents an attempted translation of the networked city ideal, in practice, it does not conform with the hegemonic premise of a networked city to foster universal water supply in the cities of the Global South.

Keywords Networked City Ideal · Urban Planning · Post-settlements Network · Water Infrastructure · WA · Ghana

---

F. Dakyaga (✉)
Faculty of Spatial Planning, Technische Universität Dortmund, Dortmund, Germany
e-mail: francis.dakyaga@tu-dortmund.de

Department of Local Governance and City Management, SD Dombo
University of Business and Integrated Development Studies, Wa, Ghana

Department of Urban and Regional Planning (URP),
Ardhi University, Dar es Salaam, Tanzania

G. Alem Gebregiorgis et al. (eds.), *Planning Cities in Africa*, The Urban Book Series,
https://doi.org/10.1007/978-3-031-06550-7_11

11.1 Introduction

Since the nineteenth century, the centralised water infrastructure has evolved as the primary archetype of residential water supply (Domènech 2011). It has been described as large-scale centralised water infrastructure (Domènech 2011), the modern infrastructural ideal (Graham and Marvin 2001), the "networked city/infrastructural ideal" (Courtard and Rutherford 2016), the "Universal access approach" (Cirolia and Rode 2019) and the modernist infrastructural ideal (Graham and Marvin 2001; Kundu and Chatterjee 2021). It gained acceptance in the mid-twentieth century as the official and universal access infrastructure doxa of modernism and comprehensive planning of public utilities in the West (Coutard 2002). In urban studies scholarship, it represents the ideal infrastructure for standardised service provision especially in cities (Cirolia and Rode 2019). In the context of this study, I use the word "networked city ideal" to showcase the provision of a large-scale water network to Wa, a secondary city of Ghana, that had been served by small-scale off-grid water systems. Since the nineties, proponents of the networked city ideal have argued for the provision of a uniform and universal water network before the growth of human settlements (Domènech 2011; Kundu and Chatterjee 2021). These proponents favour the pre-installation and provision of an engineered network of systems, fitted with technologies to facilitate the transportation of highly treated water to residents in urban areas from distant catchments (Arora et al. 2015). The principles of a networked city also include routine water purification, water as a merit or public good, and overall, the goal of universal access (Pleik 2000; Anand 2017; Pilo' 2021). Some scholars associate the public good and universal access notions of the networked city ideal, with the notion of inclusive urbanism and the right to the city (Swilling 2012; Horn 2018). Moreover, in the recent past, studies concerning southern urbanism highlight the limitations of the networked city ideal in the cities of the Global South. These studies contend that the network city ideal is more or less an imposed archetype which gained circulation in the cities of the Global South through colonialism and formal planning education (Cirolia 2020; Truelove 2019; Schramm and Wright-Contreras 2017; Sapkota et al. 2015). Some scholars suggest that engineers, planners and public health officials are key actors that have instigated the circulation or travels of the networked city ideal lured by urban modernity, and "cityness" (Monstadt and Schramm 2017).

However, lessons from "technological translations" studies suggest that translating a globalised ideal such as the "networked city ideal" remains a complex task in the local environment of the Global South, due to socioeconomic disparities between the Global South and North and among residents in the cities of the Global South (Monstadt and Schramm 2017; Kundu and Chatterjee 2020). They reveal that contradictions inherent in the implementation of the networked city ideal, reinforce the complexities. Such contradictions tend to foster uneven distribution and access to water for different socioeconomic groups across the geographies of cities (Coutard 2015; Cirolia et al. 2021). In most cities of the Global South, especially Sub-Saharan African cities, urban water infrastructure/networks commonly "follow the people", often as a post-settlement network, where settlements evolve and are more or less followed by the provision of the large-scale water network (Andreasen and Møller-Jensen 2016). This approach to water network planning may be divergent to the networked city ideal. According to Monstadt and Schramm (2017), the networked city ideal holds several interrelated notions which tend to heighten the complexities of its translation in the Global South context. It requires that; (i) a system of technology and an ordered design of cities to enable even network distribution. This is based on the notion that land-use planning is connected to universal water network provision; (ii) the notion of state monopoly—spear-headed by public utilities; (iii) it also assumes that customers are passive consumers who will eagerly wait for services to be provided; and (iv) and lastly, it also presumes that state-regulated services may offer the best results.

Observing the experiences of network water provision, Cirolia et al. (2021) indicate that the homogenous assumption of the networked city does not manifest in the cities of Sub-Saharan Africa. The networked city assumptions are less connected to the circumstances of most cities in the Global South to suit its translation. Land uses are less ordered in cities of Sub-Saharan Africa. Furthermore, residents/customers are not necessarily passive consumers but often govern themselves either by an alternative system of water supply or by seeking alternative ways of accessing water in the absence of a water network (Dakyaga et al. 2018, 2020). Consequently, the network water infrastructures often occur as post-settlement networks. Such infrastructures are provided to already developed settlements (Andreasen and Møller-Jensen 2016). Apart from Monstadt and Schramm (2017), who explore how the networked city ideal has been translated in the city of Dar es Salaam (Tanzania), studies are scant in the Global South context, on the complexities, the actors involved, the motives and processes of translating the networked city ideal, and how the actors navigate the complexities towards the hegemonic goal of universal water access.

In the literature, studies exist on the post-network infrastructure concerning the alternative infrastructural arrangements that evolved through residents' self-initiative, shaped by the failure of the large-scale or network water infrastructure (Cirolia 2020; Cirolia et al. 2021; Cirolia and Rode 2019). These studies have demonstrated how the post-network water infrastructure represents a heterogeneous infrastructural turn, an evolution as an alternative, normative ideal. The heterogeneity of infrastructure is commonly observed in water and sanitation domains, where the self-initiatives and ordinary practices of the citizenry predominate (Devlin 2018; Kundu and Chatterjee 2020; Truelove 2019; Peloso and Morinville 2014; Rusca and Schwartz 2017). Some studies also indicate that the post-settlement water network is more or less a response to the infrastructural gap left unattended to by the state in the Global South (Lawhon et al. 2018; Dakyaga et al. 2018). The aims of this study are; (i) to explore the complexities, the motives and processes of post-settlement water infrastructure planning, the actors involved; and (ii) and how they navigate the complexities towards the goal of universal water access. The next section presents how the networked city ideal manifests generally as a post-settlement network for planning in African cities.

11.2 Post-Settlement Water Network Planning in Urban Sub-Saharan Africa

In recent times, urban infrastructural studies more or less revolve around heterogeneous infrastructure, everyday practices and everyday governance debates (Velzeboer et al. 2017; Truelove 2019; Hackenbroch and Hossain 2012; Neves Alves 2019; Cornea et al. 2016). These studies reveal how urban citizens engage in self-initiatives both by providing and accessing water services in a truncated infrastructural landscape (Peloso and Morinville 2014; Uwayezu and de Vries 2020; Fakere et al. 2018; Ledo Espinoza 2020; Allen et al. 2017; Hofmann 2020; Dakyaga et al. 2020; Uitermark and Tieleman 2021). Anand highlights how citizens control, maintain and manage city water on an everyday basis even within the large-scale water network (Anand 2017). In scholarly studies of the everyday practices and governance, the micro-politics, and the roles played, the persistence and resistance of the urban citizens towards influencing institutional change and infrastructural provision have been demonstrated (Wamuchiru 2017; Mapunda et al. 2018). Lemanski (2019a) introduces the term "infrastructural citizenship" to showcase how citizens and state expectations and actions relate and shape infrastructure provision. Her findings showcase some of the characteristics of the post-settlement network water infrastructure provisioning debates in the cities of the Global South. This chapter aims to contribute to this evolving debate, particularly the work of Monstadt and Schramm on global technological circulation and translation, and the work of Andreasen and Møller-Jensen on the post-settlement water network provision; both in the context of the cities of the Global South.

Post-settlement water networks involve the provision or extension of the water network or infrastructure, either by the citizens or the utility, to areas or settlements already inhabited by residents. Post-settlement networks and services include water services provided by both large-scale and small-scale water networks. However, this study focuses on large-scale water network services at the city scale. Post-settlement water networks occur in both planned and unplanned settlements of the cities of the Global South. In Dar es Salaam, Andreasen and Møller-Jensen report that in areas already inhabited by citizens, the state utility provides electricity, public transport services, access roads, health and educational services. Due to the absence of pro-active network planning, residents often live in unplanned areas with the anticipation of future network service provision. Utility services are provided years after the development of settlements. Residents access services through negotiation with formal institutional actors, submitting formal applications to the utility for network extension and sometimes co-financing the cost of the water provision (Andreasen and Møller-Jensen 2016). Post-settlement water network provision represents a formal and a reactive planning approach to infrastructure provision. Also, post-settlement water network planning is often vulnerable to politics. Residents with the financial capacity get adequate access to the utility, unlike the lower-income residents. According to Andresen and Møller-Jensen, the politicisation of post-settlement water network provision tends to deepen infrastructure access inequalities in areas already inhabited by residents. These studies recommend further research to scrutinise the challenges surrounding post-settlement water infrastructure planning (Andreasen and Møller-Jensen 2016). Similarly, in Accra (Ghana), Uitermark and Tieleman (2021) report that residents lived in peripheral settlements for three years, without network service such as water, electricity, access roads and drainage. The residents gained access to utility services and infrastructure through co-financing.

Similarly, in the city of Addis Ababa (Ethiopia), Cirolia et al. (2021) find that the state fills infrastructural gaps within the fragmented systems through the provision of high-tech sanitation infrastructure in residential apartments or condominiums already inhabited by residents. Such attempts to fill infrastructure gaps through the provision of large-scale infrastructure tend to occur in settlements where residents have lived for decades. This practice contributes to infrastructure fragmentation (Simonovic and Arunkumar 2016). Post-settlement water networks also occur where both the large-scale water infrastructure and the alternative water systems are provided in areas already inhabited by residents. Both arrangements may conflict with the networked city ideal and contribute to the water infrastructure fragmentation in African cities (Simonovic and Arunkumar 2016).

Also, Cirolia and Rode (2019) advance the potential of the small-scale post-settlement water network. In their theorisation of post-network water and the new Sufficiency Approach to infrastructure provision in the Global South, they argue that the small-scale off-grid water infrastructures are often steered by the urban citizens. However, both systems may co-exist in cities as complementary networks (Kjellén 2000). In the city of Dar es Salaam (Tanzania), Andreasen and Møller-Jensen (2016) observe how residents living in peripheral areas self-organise, financing the provision of alternative water infrastructures as immediate coping measures, and yet negotiate with the water utility for the extension of the network water. The provision of the network water occurs at the latter stage in areas where residents have lived many years (Nganyanyuka et al. 2015). The provision of both large and small-scale water network via the post-settlement network conflicts with the principles of the network city ideal, particularly, the principle of pro-active service provision (Uitermark and Tieleman 2021). However, post-network water infrastructure provision may be vulnerable to high cost, conflict in land acquisition processes and unequal access to water. To limit these challenges, Andreasen and Møller-Jensen (2016) recommend a communal, coordinative and facilitative approach to post-settlement water network provisioning, beyond the sole powers of the state utility actors.

Further studies reveal that a post-settlement water network requires differentiated levels of investments (adequate financial ability) to foster equitable access in cities (Cirolia and Rode 2019). For instance, residents located farther away from the network may bear a greater cost of connection in comparison to residents located closer to the distribution network. In response, Cirolia (2020) explains that infrastructure financing requires huge financial investment; unfortunately, the financial capacities of the utilities in Africa are already in a fractured state. In the case of Africa, the fractured fiscal authority and delays in negotiating for funding are the reasons for the evolution of the post-settlement water network. Cirolia and others further note that due to fiscal restraint, the financialisation of network water infrastructure is often augmented by development agencies, non-government organisations and civil society organisations (Peter-Varbanets et al. 2009; Cirolia 2020). The delays in network water provision may be caused by the long-term arrangements in accessing donor funding and loaning facility for large-scale water infrastructure development (Banerjee et al. 2009).

In the context of technological translation, Monstadt and Schramm (2017) suggest that, in part, the post-settlement water networks represent state's attempt to provide network water, and on the other hand failures in an attempt to foster universal water access. In the city of Dar es Salaam, Monstadt and Schramm (2017) find that the translation of the globalised networked city ideal involved adaptation and some level of creativity from the citizens. In addition, the creativity, capital investment capacity and the technical and managerial abilities of the actors involved are fundamental for a comprehensive translation of the networked city ideal (Wamuchiru 2017). Collier and Venables note that the cost of extending the water network is often three times higher in sprawling cities and low-density areas than in cities characterised by high-density (Collier and Venables 2016). As a result, African cities may require more financial resources to enable universal access to water through the translation of the networked city ideal. Andreasen and Møller-Jensen (2016) explain that post-settlement water network extension may be seen more often than pre-settlement network extension because the former occurs in sprawling cities, where settlements are often not in conformity with spatial planning regulations. Though the majority of the residents often dwell in the high-density areas, (the inner cities zones), a lack of spatial order challenges water network extension (Mcgranahan et al. 2020). Chaotic growth of inner-city zones acts as a barrier to even distribution and access to network water (Guma and Schramm 2019). Eggimann and others explain that the uneven distribution of water service exists within the networked city because, it is economically inadvisable to provide network water infrastructure to low-density areas (Eggimann et al. 2016). Also, Ranganathan (2005) and Lemanski (2019b) use the concept of "urban citizenship" to explain that the uneven distribution of a water network occurs because of the utility redefinition of citizenship. Urban citizenships occur where certain legal documents are requested by the utility from potential customers and residents for proof of identity for connectivity to the water network. Such documents may include the land title deed, the electricity bill and the site plan as a means of identifying potential water users. They argue that the redefinition of citizenship as "urban citizenship" often fosters inclusion and exclusion in terms of residents' access to the utility network.

Conversely, these debates suggest that urban infrastructure provision ought to be a shared task that can be collectively understood and provided by varied actors (Wamuchiru 2017; Truelove and Cornea 2020). However, though studies exist on the financialisation of post-settlement water networks, there are few studies on the complexities, the actors, the motives and processes, as well as how the actors navigate towards attaining the goal of universal water access in the Global South.

11.3 Study Context and Methods

11.3.1 Wa: A Secondary City in Ghana under Network Water Transition

Wa is a secondary city and the administrative capital of the Upper West Region (UWR) of Ghana. Located in the semi-arid region of north-western Ghana, Wa is one of the least urbanised secondary cities, but has been urbanising rapidly in recent times (Oteng-Ababio et al. 2019). In 2012, the urban population growth rate stood at 4 per cent, higher than the national rate of 3.4 per cent (Osumanu et al. 2018). In 1880, the city had a population of approximately 8000 (Wilks 1989) which rose to 125,479 by 2017. Like other cities of Ghana, Wa had experienced inadequate access to potable water for many years. In Ghana, the dominant system of water supply was the groundwater (boreholes and wells) until 1928, when the network water supply was instituted and operated by the Hydraulic Division of the Public Works Department (PWDs).

The department was first established in Accra in what was then the Gold Coast and tasked with urban water supply. Even then, the networked water system was primarily provided to the residences of the British, the colonial masters (Twumasi 2013), see Fig. 11.1

In the 1880s, the residents of Wa depended on groundwater systems such as hand dug-outs wells and dams. Finally, in the 1980s, the Hydraulics Division of the Public Works Departments (PWDs), supplied the main source of water to residents through three drilled mechanised boreholes. Pipelines were laid within the township to facilitate the distribution of the water from the boreholes. The boreholes were connected to pump machines and a treatment system located in the centre of the city. The water treatment tank had a storage capacity of 40,000 gallons2. As the population increased the capacities of the storage tanks and the boreholes were inadequate. The water utility authorities provided standpipes in major areas of the city connected to the water storage tanks to facilitate access. Furthermore, in 1984, additional treatment and water storage capacity was required to serve the population of 36,000. A water treatment plant with a storage capacity of 1800 m3, and a daily production capacity of 1320 m3, was provided to increase water access (Twumasi 2013). In the 1950s, a total of 16 boreholes were drilled, mechanised for water supply. By 2010, the population of Wa rose to 71,340. Residents were supplied water from 18 drilled and mechanised boreholes. In 2017, the population grew to 125,479 (Ghana Statistical Service 2014; Ahmed et al. 2020).

As of 2017, the water network covers only the townships, to the exclusion of residents in the extreme peripheral areas of the city. The study used the qualitative case study research method to explore the motives, processes and complexities of post-settlement water network planning, the actors connected to the process and how they navigate the complexities towards universal water access. This method was found advantageous for unpacking the complexities that characterise post-settlement water network planning in a secondary city such as Wa. To gain a clear understanding

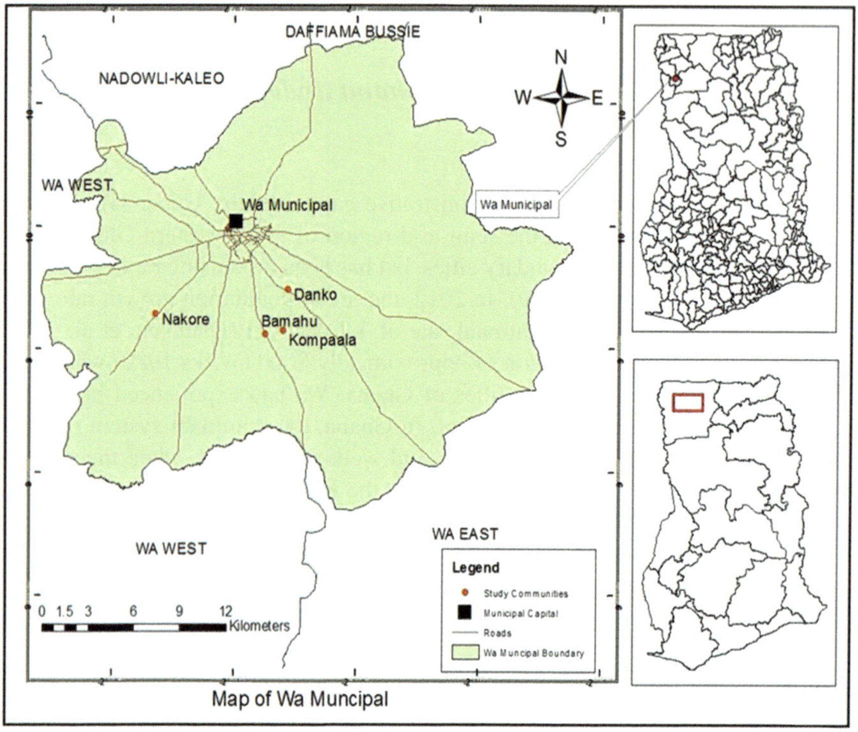

Fig. 11.1 Map of Wa township (*Source* Author's construct)

of the complexities across the city in terms of network water provision, the city was divided into three spatial zones; see Fig. 11.2.

The spatial zones are: the core urban area/ Central Business District (CBD), the transitional zone; and the peripheral zone. Two settlements were purposively selected from each of the zones. In the core urban zone/ CBD, Limanyiri/ Fifamuni and Dokpong were selected. In the transitional zone of the city, Degu residential area and Konta were selected. Whilst in the peripheral zone, Mangu and Dobile were selected. These zones were selected with the support of the physical planning unit of the municipality.

Semi-structured interviews were conducted with a total of sixteen participants, who were key sources of information, directly involved in the city network water provision. The interviews took the form of face-to-face discussions using a semi-structured interview guide. The participants comprised of the Water Distribution Engineer of the Ghana Water Company Limited (GWCL), the state utility, the Water Quality Manager and the Director of the utility, the Physical Planner, Development Planner, Manager of the Water Resources Commission and the local political representatives (Assembly members) of the selected residential zones. These participants

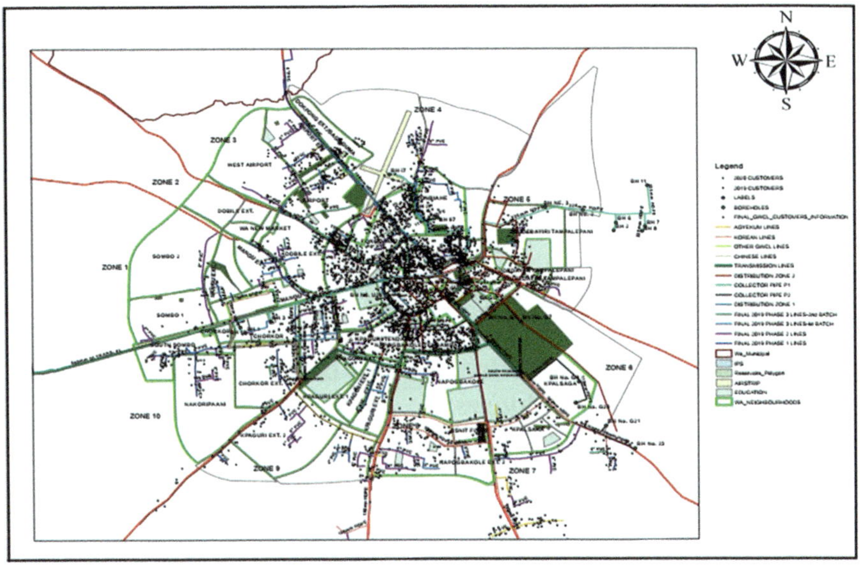

Fig. 11.2 Map showing networked pipeline (*Source* Ghana Water Company Limited, GIS Lab)

were selected based on their experience and knowledge of the network water extension in the city. Focus Group Discussions (FGD) were also conducted with six groups of residents, one from each of the six selected urban communities. The discussion groups ranged in size from 7–10 participants each.

In both the interviews and the FGDs, the study probed into the processes involved in network water provision in already existing urban settlements and the actors engaged in the water extension. Also, given the existence of the off-grid water systems (boreholes and dug-wells) in the city, before the provision of the networked water, the study investigated, whether there was a need for networked water infrastructure the actors involved and the complexities that arose through the provision of the post-settlement water network. And lastly, I questioned the state utility actors (Engineers, Water Quality Manager, and Director) as to whether there were possibilities of attaining universal access to water via the post-settlements network.

The utility officials described the processes that guided the network water extension. In addition, transect walks were conducted and guided by the Municipal Director of the utility and the Assembly members (local political representatives) of the selected communities. The transect walks led to observations on the nature of the settlements in the three urban zones, the pipe networks and other alternative water systems in the various communities. The interviews and discussions enabled the collection of data on the processes and the actors linked to the post-settlement network water provision, to understand the complexities and how the actors involved react to the challenges in network water extension. The interviews and the FGDs sessions were recorded. The interview sessions lasted from between 45 min to an hour. The audio recordings were transcribed and analysed thematically, taking into consideration the different processes, and descriptions obtained from the respondents. The data were collected between December 2020 and August 2021. An analysis was done across the various spatial zones in terms of water access, and the diverse complexities that arose in the attempt to network water, and how both the utility providers and consumers navigated the process of the network extension.

11.4 Results and Discussions

11.4.1 Post-Settlement Water Network Extensions: The Actors and the Motives

Consistent with Cirolia (2020), observations of Sub-Saharan African cities where fractured financial capacity limits network extension, the study found that the water treatment plant for water supply in the secondary city of Wa was financed by a loan facility acquired from the Korean EXIM Bank. The loan facility amounted to $51.54 million, acquired by the state. In line with the networked city ideal of Courtyard and Rutherford (2015), the motive of the large-scale water infrastructure was neither to maximise profit nor mobilise revenue and recover the cost, but for the utility to break-even, and increase networked coverage towards universal water access. In this context, the provision of network water became necessary following the rise in the population of the city and the inability of the utility to achieve adequate water supply coverage with the use of the small-scale off-grid, mechanised water systems (groundwater from boreholes). With the rise in the population of the city from 71,340 in 2010 to 125,479 in 2017, only 12 per cent of the city's population could be served water from the 18 mechanised boreholes. The utility established the water treatment plant as a response to the water supply deficit. The treatment and distribution plant of the city water network was established in Jambussi, a community located about 35 kms away from the township of Wa. The trenches were sunk and the major transmission pipelines laid from the main reservoir to the city. The major transmission pipelines cut across several rural communities and arable lands. The high cost of financing the post-settlement water network was due to the cost of extending the network to peri-urban areas and the cost of compensating residents directly or indirectly affected by the network water extension (Andreasen and Møller-Jensen 2016). The utility compensated resisting and agitating resident farmers who were affected by the network extension. The damages caused to farmlands through the extension of the main water distribution pipelines were assessed and the affected residents compensated through direct cash payment. Even when, the affected arable lands were undeveloped plots, the utility (GWCL) paid about GHC 32,000, ($5,280.60) as compensation to the affected farmers in 2020. These petty costs tend to increase the cost of financing post-settlement water infrastructure (Collier and Venables 2016). The utility paid the compensation, drawing monies from its financial operations.

Moreover, residents of the rural communities were not connected to the networked water, as they were satisfied with borehole water systems at the time. However, with the recent growth of the population, residents in the affected villages have begun agitating for a network water supply. The utility has remained mute, as they fear the defaulting of water bills.

Multiple actors, including state and non-state actors, contributed to the development and the expansion of the network water infrastructure in a collaborative manner (Andreasen and Møller-Jensen 2016). Among them were spatial planners, development planners from the municipality, Lands Commission officials, the Water Engineer in charge of the network distribution, the Department of Urban Roads and of telecommunications, as well as Assembly members (local political representatives), traditional leaders, local plumbers, water vendors, tanker drivers and community water board managers. The major state actors were from the management of the water utility. The treatment plant was developed by a hired contractor, with overarching supervision, operation by the management of the state's water utility. Though the settlements had sprawled to about 24.6 km, with a population density of 3,253 people per square kilometre as of 2016 (Osumanu et al. 2018), the state's water utility was confident in its ability to provide water to the doors-steps of all of the households in the existing city. Individual plots and house owners were required to make a formal request for a site plan from the Land Use and Spatial Planning Authority (LUPSA). Through the LUPSA, site plans were provided to homeowners to facilitate the identification of utility spaces in the settlements that were already developed. In this way, the state utility redefined urban citizenship that offers access to network water. Ownership or possession of a site plan, and electricity bills led to eligibility or non-eligibility to networked water (Ranganathan and Balazs 2015; Lemanski 2019b). The eligibility or non-eligibility were also determined based on land-use order (Monstadt and Schramm 2017). Due to uncontrolled urbanisation (Osumanu and Akomgbangre 2020), most parts of the inner city were unplanned, hence lacked a site plan to guide the extension of the water network. The development planners at the City Municipality provided directives and guidance on areas suitable for water utility networked extension. The process encountered some resistance from residents (Lemanski 2019b). The resistance occurred due to encroachment of the utility networked on individual plots, and the payments of the connection fees. Despite the resistance, the post-settlement water provision contributed to (un)making the city and infrastructure. The Survey Department of the Land Commission, the traditional leaders and the Assembly members supported the state utility in networking the city, by negotiating with the residents and landowners who were affected through the laying of pipes across their undeveloped plots. The state water utility distribution engineer was tasked with an aggressive extension to the water network. Contractors were hired to facilitate the extension through the sinking of trenches within settlements and the building of long roads for the network extension. In the first phase in 2019, the utility created about 18 square kilometres of coverage with the main distribution pipelines, and subsequently covered about 28 square kilometres with pipelines for household connections. The local plumbers were hired to sink trenches and lay pipelines for residents who paid for the utility connection.

11.4.2 Post-Settlement Water Network Extension: Criteria and Processes to Access the Network

Consistent with related studies of formal water utility arrangements, the criteria and processes for network water provision were based on formal application processes for a utility connection (Andreasen and Møller-Jensen 2016) defined by identity/citizenship with emphasis on the possession of legal documents, references and identification (Ranganathan 2005). As part of formal applications by homeowners for household water connections, the state water utility's first eligibility requirement was for a site plan of the residence to be provided. In addition, a utility bill, for example, an electricity bill, was also required to further verify and affirm the urban citizenship of residents (Lemanski 2019b). As seen in Lemanski's (2019b) and Ranganathan's (2005) perspectives of citizenship, the request for the site plan from potential clients of the utility served as an instrument to identify clients, and also symbolises an indirect inclusion and exclusion criteria, because only residents with site plan of the dwelling could access utility running water. Also, the use of the site plan by the state utility as a requirement for water network extension represents a translation of the networked city ideologies, in which urban areas with ordered land uses are given preference through the provision of network water (Monstadt and Schramm 2017). The site plan thereby provided proof of the location of the plot and residence. Moreover, an in-depth interview with the spatial planner revealed that the use of individual site plans by the utility as a direction for water network extension can be misleading in the context of Global South cities, because an individual plan does not reveal the entire characteristics of the individual neighbourhood plan. But for the state water utility, the individual household site plan was an important guide for the post-settlement water network extension process. In addition, for adequate connectivity and reliable access to water, the distance of a clients' household from the main transmission lines/pipes could not exceed a maximum of 200 m. Though it was essential to ensure that residents received reliable water flow when connected, the utility policy of 200 m as the standard distance for a potential water network, resulted in the exclusion of residents located beyond the 200 m distance. The location of the plots and the type of occupancy of the property, for example, tenant, landlord, free occupant, or family client, also determined utility network water access and connection. The utility distribution team conducted an estimate of the cost of materials, by measuring the distance between the main transmission pipelines to the plots of the clients who had met the connection requirements.

Ultimately, connectivity to the utility network depended on the ability of the household to pay for the estimated cost of the materials provided by the state water utility. The interviews also revealed that the lowest estimated cost for connection at the shortest distance was about GHC 800.00, ($133.00). Whereas the highest estimated cost of connection for residents located at a distance of approximately 200 m was about GHC 2,500.00 ($398.78). Heads of households and landlords bore the cost of network water extension. This corroborates with Andreasen and Møller-Jensen's (2016) observations that residents beyond the networked made formal applications for post-networked services. Findings of Uitermark and Tieleman (2021) in Accra, where residents contributed to the co-financing of the extension of networked water after years without the network's water is consistent with the findings of this study. Moreover, the cost was often uneven, the estimated cost varied across locations of households and depended more on the distance of households from the transmission pipelines. The closer the household to utility transmission pipelines, the lower the estimated cost of materials for network connection. Residents in the transitional zone of the city were at a greater advantage due to ordered land uses, especially networked roads, compared with residents who dwelled at the periphery and in the inner city. Exceptionally, most residents in the inner city were indirectly excluded, due to congestion, lack of space for utility network connection and disordered land uses. Whilst residents at the periphery were not connected either due to deliberate refusal to bear the estimated cost, or inability to access the network due to their location beyond the utility limited connectivity standard distance of 200 m. The interviews also revealed that most high and middle-income residents, especially those who owned houses, were more interested in getting connected to the utility network compared with those in rental housing units. Especially, landlords (rental house owners), who were already connected to the small-scale off-grid mechanised water systems were not interested in connecting to the network, because their tenants served as customers of their drilled mechanised water systems. Therefore, networking their homes with the state water network would reduce the income they could gain income through the sale of the drilled groundwater to their tenants, who served as their customers.

11.4.3 Post-Settlement Water Networking: The Growing Complexities and the Utility Navigations

Post-settlement water network extension is often prone to several complexities. The first major difficulty arose in the distribution/extension of transmission and service pipelines due to disordered land use and the sprawling growth of the city. The interviews revealed that the extension of the pipelines for transmission and services was challenged by the disordered and sprawling growth of the city, especially congestion and limited space for the utility network in the inner-city zone. This challenged the even distribution of the transmission lines for universal water access for households' connectivity. Other extension challenges included the inability of most residents to afford the estimated cost of materials to enable connection to the state utility network (Coutard 2015; Cirolia et al. 2021). Also, land disputes arose due to the wrongful provision of standpipes on undeveloped plots/spaces owned by residents. In accordance with Monstadt and Schramm's (2017) observations of the creativity and adaptation of translated technologies, the study found that the utility officials navigated the issues of congestion and the disordered growth of the city by inventing multiple systems of water supply within the networked city. They included the following;

firstly, the provision of direct water networked to households that were within the standard distance of 200 m and could afford the estimated cost of materials for connection. This included some residents of the inner cities. In some areas of the inner city, the footpaths of residents were used as service spaces/paths. Trenches were sunk and pipelines laid using the footpaths that led residents to their homes due to the lack of space for a utility network. The use of the narrow footpaths as utility and services' spaces was an innovative and creative idea which prevented the utility from demolishing haphazard structures. The use of the footpaths as spaces for services helped the utility to avoid the cost of compensating affected residents. The innovation by the utility officials towards creating universal water access was the provision of standpipes in available spaces of the inner city (Dakyaga et al. 2018). It evolved as an alternative to reduce the time residents spent collecting water and the distance they travelled to get it. Areas with low utility networks and low financial capacity to pay for the estimated cost of the utility network connection were provided standpipes as an alternative water supply model. The utility bore the cost of the standpipe connections. Selected residents, mostly women, acted as standpipe vendors and were paid monthly commission for the sales of water. The third innovation by the utility to enable water access was the use of utility tanker trucks. The tanker trucks were an alternative means of reaching residents who either lived beyond the network or were excluded by the utility's standard distance of 200 m from the transmission pipes. The tank trucks collected water from the utility and delivered it to residents in response to clients' request. These modes of water service delivery confirm the perspective of Cirolia et al. (2021) on the heterogeneity of infrastructure within the networked city, that commonly exists in the cities of the Global South, in contrast to the premise of the homogeneity of the networked city ideal.

The second major complexity relates to technical and operational issues. This involves the intermittent disruptions to the water flow of the already connected households due to repairs and maintenance and ongoing network extension by the utility. During the interruption to the water flow, the state water utility relied on the small-scale off-grid systems previously used by the utility as the main source of supply water in the city. The reliance on the alternative small-scale off-grid water systems existed as a complementary system to balance the water available and to aid transition into the networked city ideal (Cirolia 2020). Moreover, the small-scale off-grid water systems seldom serve the entire city during a period of interruption. Another technical and operational challenge found was the utility's lack of adequate financial capacity to purchase transmission pipelines to increase the coverage of the water network. This challenge was compounded by the increasing water bill delinquency by connected residents and state institutions. The interviews revealed that delinquent bills arose not only due to the inability of residents to pay but also because payment of water bills represented a new urbanism. The utility addressed these complexities using multiple tactics. Firstly, by engaging in an aggressive door-to-door campaign confronting delinquent customers for water bill collection. The commercial team of the utility undertook periodic outreach projects with the use of a truck van and client profiles to mobilise payment of water user fees. However, this tactic was only effective when the commercial outreach team met clients at their homes. In this context, the identity of the clients represented not only "urban citizenship" by right to water access (in)exclusion from the network (Ranganathan and Balazs 2015), but also a mechanism and tactic for user fee mobilisation in an era of defaults. The second tactic was the disconnection of defaulters from the utility network. Disconnection tends to increase the cost to the clients, because disconnected customers/clients were compelled to pay an amount of money, in addition to unpaid bills for re-connection to the utility. Moreover, disconnection became necessary when routine visits by the utility's commercial team to the clients' homes for water bills mobilisation proved futile. Interviews revealed that disconnection was a less effective measure in encouraging the payment of water bills among low-income residents. However, it was found effective among middle and high-income residents living in gated houses who were less interested in carrying buckets of collected water from a far distance. For most low-income residents, the act of paying a bill for water already consumed and monies for the cost of re-connection to the utility were non-incentives. Instead, most low-income residents disconnected from the network of the state utility relying on the small-scale off-grid water systems for water. In this context, the existence of alternative water systems acted as a barrier to the universal networking of the city. Interviews revealed that it took about three to fourth months for most disconnected clients to visit the utility office to express interest in re-connection to the utility network. The utility lamented that the existing alternative systems of water supply limit attaining universal water network extension. Thirdly, the utility developed the tactics of inviting defaulters to their office for deliberation, discussions and negotiation on bill payments. Through the negotiation, the utility provided debt payment schedules that were commensurate with the financial capacity of defaulters. Defaulters who resisted the payment negotiations with the utility officials were sent to court to serve as a deterrent to other clients.

11.5 Conclusion

The study explored the actors involved, the motives, processes and complexities of post-settlement network extension and how the actors navigated the complexities towards attaining the goal of universal water access through the secondary of Wa, in Ghana. A need was seen for this due to limited studies on post-settlement water networking and the complexities that characterise this kind of infrastructure planning by water utilities in the cities of the Global South. The study discovered that though the utility providers were motivated to achieve citywide coverage of water network connectivity through a post-settlement water network in the image of the networked city, universalising access was a critical challenge. The challenges arose due to resistance of the inhabitants, the disordered spatial expansion and the sprawling growth of the city. The sprawling growth of the city required that the utility invest more financial capital to purchase pipelines to expand coverage of the network, to ease and lower the estimated cost of materials to enable low-income households to connect to the network. The translation of the networked city created spatial segregation due to access advantage and disadvantage within the city's zones and among the residents. As a result, residents in the intermediary city or the transitional zone were the most served, due to their proximity to the main transmission pipelines and their ability to afford the estimated cost of the materials. The utility navigated these complexities by inventing and devising some creative solutions to fulfil the principle of the networked city ideal (universal water access). Instead of the mono-modal/homogeneous system, the utility invented heterogeneous water supply models within the networked city ideal. This entailed the provision of indirect and direct water supply systems as well as the transportation of water via tanker trucks to unserved residents. The heterogeneity of supply models arose through creativity and the aim of the utility to achieve universal water access. In addition, the indebtedness of residents and state institutions challenged the financial capacity of the utility in improving coverage. For most low-income residents, payments to the utility were seen as warranted by the residents at the time water was required for consumption, but not after consumption. As result, most low-income residents defaulted on the payment of utility bills. This tends to challenge the utility's financial capacity to expand coverage towards universal water access. The post-settlement water network represented the attempted translation of the networked infrastructural ideal from the Global North to the Global South. But, in fact, the translation of the networked city ideal in the Global South context, often encounters several complexities requiring the need for adaptation, innovation and creativity.

The roles of small-scale off-grid systems were crucial and complementary in enabling the transition of residents from the small-scale off-grid systems towards the networked city ideal—a large-scale centralised water infrastructure. Moreover, residents and the state utility remained connected to the small-scale off-grid water systems, such as the boreholes fitted with hand pumps, mechanised boreholes and the protected wells for use during the intermittent disruption of the networked water for repairs, maintenance and the connection of new clients to the network. Overall, post-settlement water network extension has the potential for universalising water access in ordered cities, in which socioeconomic inequalities among the residents are minimal. However, collaboration among the varied actors is key in advancing universal water access through post-settlement water infrastructure planning.

References

Ahmed A, Korah PI, Dongzagla A, Nunbogu AM, Niminga-Beka R, Kuusaana ED, Abubakari Z (2020) City profile: Wa, Ghana. Cities 97

Allen A, Hofmann P, Mukherjee J, Walnycki A (2017) Water trajectories through non-networked infrastructure: insights from peri-urban Dar Es Salaam, Cochabamba and Kolkata. Urban Res Pract 10(1):22–42

Anand N (2017) Hydraulic city water and the lifrastructures of citizenship in Mumbai. Duke University Press, Durham and London

Andreasen MH, Møller-Jensen L (2016) Beyond the networks: self-help services and post-settlement network extensions in the periphery of Dar Es Salaam. Habitat Int 53:39–47

Arora M, Malano H, Davidson B, Nelson R, George B (2015) Interactions between centralized and decentralized water systems in urban context: a review. Wiley Interdiscip Rev Water 2(6):623–634
Banerjee S, Wodon Q, Diallo Q, Pushak T, Uddin E, Tsimpo C, Foster V (2009). Access, affordability, and alternatives: modern infrastructure services in Africa
Cirolia L, Rode P (2019) Urban infrastructure and development. LSE Cities Working Papers, 18. https://lsecities.net/wp-content/uploads/2019/01/GII-Working-Paper-1-Urban-infrastructure-development.pdf
Cirolia LR (2020) Fractured fiscal authority and fragmented infrastructures: financing sustainable urban development in sub-saharan Africa. Habitat Int 104:102233
Cirolia LR, Hailu T, King J, da Cruz NF, Beall J (2021) Infrastructure governance in the post-networked city: state-led, high-tech sanitation in Addis Ababa's condominium housing. Environ Plan C: Polit Space 39(7)
Collier P, Venables AJ (2016) Urban infrastructure for development. Oxf Rev Econ Policy 32(3):391–409
Cornea N, Zimmer A, Véron R (2016) Ponds, power and institutions: the everyday governance of accessing urban water bodies in a small Bengali city. Int J Urban Reg Res 40(2):395–409
Coutard O (2002) 'Premium network spaces': a comment. Int J Urban Reg Res 26.1:175–182
Coutard O (2015) Beyond the networked city. https://doi.org/10.4324/9781315757612
Courtard O, Rutherford, J (2016) Beyond the Networked City: Infrastructure reconfigurations and urban change in the North and South. Routledge, New York
Dakyaga F, Ahmed A, Sillim ML (2020) Governing ourselves for sustainability: everyday ingenuities in the governance of water infrastructure in the informal settlements of Dar Es Salaam. Urban Forum 32:111–129
Dakyaga F, Kyessi AG, Msami JM (2018) Water access today and tomorrow: domestic water sustainability under informal water supply markets in Dar Es Salaam, Tanzania. J Sustain Dev 11(6):120
Devlin RT (2018) Asking 'third world questions' of first world informality: using southern theory to parse needs from desires in an analysis of informal urbanism of the global north. Plan Theory 17(4):568–587
Domènech L (2011) Rethinking water management: from centralised to decentralised water supply and sanitation models. Documents D'analisi Geografica 57(2):293–310
Eggimann S, Truffer B, Maurer M (2016) Economies of density for on-site waste water treatment. Water Res 101:476–489
Fakere AA, Folorunso CO, Omole FK, Ayoola HA, Arayela O (2018) An assessment of self-help water supply in Akure, Nigeria. Util Policy 51:89–94
Ghana Statistical Service (2014) Ghana living standards survey (GLSS6). Ghana, Accra
Graham S, Marvin S (2001) Splintering Urbanism: Networked infrastructures, technological mobilities and the urban condition. Routledge, New York.
Guma PK, Schramm S (2019) Hybrid constellations of water access in the digital age: the case of Jisomee Mita in Soweto-Kayole, Nairobi. Water Altern 12(2):636–654
Hackenbroch, K, Hossain S (2012) 'The organised encroachment of the powerful'-everyday practices of public space and water supply in Dhaka, Bangladesh. Plan Theory Pract 13(3):397–420
Hofmann P (2020) Meeting WASH SDG6: insights from everyday practices in Dar Es Salaam. Environ Urban. https://doi.org/10.1177/0956247820957280
Horn P (2018) Indigenous peoples, the city and inclusive urban development policies in Latin America: Lessons from Bolivia and Ecuador. Dev Policy Rev 36(4):483–501
Kjellén M (2000) Complementary water systems in Dar Es Salaam, Tanzania: the case of water vending. Int J Water Resour Dev 16(1):143–154
Kundu R, Chatterjee S (2020) Pipe dreams? Practices of everyday governance of heterogeneous configurations of water supply in Baruipur, a small town in India. Environ Plan C: Politics Space 39(1)

Kundu R, Chatterjee S (2021) Pipe dreams? practices of everyday governance of heterogeneous configurations of water supply in Baruipur, a small town in India. Environ Plan C: Politics Space 39(2):318–335
Lawhon M, Nilsson D, Silver J, Ernstson H, Lwasa S (2018) Thinking through heterogeneous infrastructure configurations. Urban Studies 55(4):720–732
Ledo Espinoza PJ (2020) Peri-urbanization in Sacaba, Bolivia: challenges to the traditional urban planning approach. Int Plan Stud 26(4):1–16
Lemanski C (2019a) Infrastructural citizenship: (De)constructing state-society relations. https://doi.org/10.17863/CAM.46880
Lemanski C (2019b) Infrastructural citizenship: the everyday citizenships of adapting and/or destroying public infrastructure in Cape Town, South Africa. Trans Inst Br Geogr 45(3):589–605
Mapunda DW, Chen SS, Yu C (2018) The role of informal small-scale water supply system in resolving drinking water shortages in peri-urban Dar Es Salaam, Tanzania. Appl Geogr 92:112–122
Mcgranahan G, Kyessi A, Andreasen MH, Kombe WJ (2020) Investigating the urban land nexus and inclusive urbanisation in Dar Es Salaam, Mwanza, and Khartoum three city land nexus research team. https://doi.org/10.13140/RG.2.2.26511.74406
Monstadt J, Schramm S (2017) Toward the networked city? Translating technological ideals and planning models in water and sanitation systems in Dar Es Salaam. Int J Urban Reg Res 41(1):104–125
Neves Alves S (2019) Everyday states and water infrastructure: insights from a small secondary city in Africa, Bafatá in Guinea-Bissau. Environ Plan C: Politics Space 39(2):247–264
Nganyanyuka K, Martinez J, Wesselink A, Lungo JH (2015) Accessing water services in Dar Es Salaam: are we counting what counts? Habitat Int 44(2014):358–366
Osumanu, IK, Akomgbangre JN (2020) A growing city: patterns and ramifications of urban change in Wa, Ghana. Spat Inf Res. https://doi.org/10.1007/s41324-020-00313-1
Osumanu IK, Akongbangre JN, Tuu GN, Owusu-Sekyere E (2018) From patches of villages to a municipality: time, space, and expansion of Wa, Ghana. Urban Forum 30:57–74
Oteng-Ababio, M, Owusu G, Asafo DM (2019) Following the footsteps: urbanisation of Wa municipality and its synergism in risk accumulation, uncertainties and complexities in urban Ghana. Jamba: J Disaster Risk Stud 11 (1):1–10
Peloso M, Morinville C (2014) 'Chasing for water': everyday practices of water access in peri-urban Ashaiman, Ghana. Water Altern 7(1):121–139
Peter-Varbanets M, Zurbrügg C, Swartz C, Pronk W (2009) Decentralized systems for potable water and the potential of membrane technology. Water Res 43(2):245–165
Pilo' F (2021) Negotiating networked infrastructural inequalities: governance, electricity access, and space in Rio de Janeiro. Environ Plan C: Politics Space 39(2):265–281
Pleik, P (2000) The changing water paradigm: a look at twenty-first century water resource development. Water Int 25(1):127–138. http://www.mendeley.com/research/the-changing-water-paradigm-a-look-at-twentyfirst-century-water-resources-development/%5Cn, http://www.unc.edu/~charack/pdf/Gleick_Article.pdf
Ranganathan M (2005) Reengineering citizenship: municipal reforms and the politics of 'e-grievance redressal' in Karnataka's cities. In: Desai R, Sanyal R (eds) Urbanizing citizenship: contested spaces in Indian cities. Sage, New Delhi and Thousand Oaks, pp 109–132
Ranganathan M, Balazs C (2015) Water marginalization at the urban fringe: environmental justice and urban political ecology across the north-south divide. Urban Geogr 36(3):403–423
Rusca M, Schwartz K (2017) The paradox of cost recovery in heterogeneous municipal water supply systems: ensuring inclusiveness or exacerbating inequalities ? Habitat Int 1–8.
Sapkota M, Arora M, Malano H, Moglia M, Sharma A, George B, Pamminger F (2015) An overview of hybrid water supply systems in the context of urban water management: challenges and opportunities. Water (switzerland) 7(1):153–174
Schramm S, Wright-Contreras L (2017) Beyond passive consumption: dis/ordering water supply and sanitation at Hanoi's urban edge. Geoforum 85:299–310

Simonovic SP, Arunkumar R (2016) Quantification of resilience to water scarcity, a dynamic measure in time and space. IAHS-AISH Proceedings and Reports 373:13–17
Swilling M (2012) Contesting inclusive urbanism in a divided city: The limits to the neoliberalisation of Cape Town's energy system. Urban Stud, 51(15):3180–3197
Truelove Y (2019) Gray zones: the everyday practices and governance of water beyond the network. Ann Am Assoc Geogr 109(6):1758–1774
Truelove Y, Cornea N (2020) Rethinking urban environmental and infrastructural governance in the everyday: perspectives from and of the global south. Environ Plan C: Politics Space 39(2):231–246
Twumasi SA (2013) Evolution of water systems and its challenges in the Wa municipal of Ghana. J Environ Earth Sci 3(7):15–24
Uitermark J, Tieleman J (2021) From fragmentation to integration and back again: the politics of water infrastructure in Accra's peripheral neighbourhoods. Trans Inst Br Geogr 46(2):347–362
Uwayezu E, de Vries WT (2020) Access to affordable houses for the low-income urban dwellers in Kigali: analysis based on sale prices. Land 9(3)
Velzeboer L, Hordijk M, Schwartz K (2017) Water is life in a life without water: power and everyday water practices in. Habitat Int 73:119–128
Wamuchiru E (2017) Beyond the networked city: situated practices of citizenship and grassroots agency in water infrastructure provision in the Chamazi settlement, Dar Es Salaam. Environ Urban 29(2):551–66
Wilks I (1989) Wa and the wala: Islam and polity in Northwestern Ghana. Cambridge, Cambridge University Press

Francis Dakyaga is an urban and regional development planner. He is a Ph.D. candidate in Urban Planning and Management of TU Dortmund, Germany and Ardhi University in Tanzania (a joint Ph.D.). He holds an MSc in Urban and Regional Development Planning and Management from TU Dortmund, Germany and Ardhi University in Tanzania, and a BA in Integrated Community Development from the University for Development Studies (UDS) in Ghana. He works as an assistant lecturer in the Department of Local Governance and City Management, SD Dombo, and at the University of Business and Integrated Development studies in Ghana. His research and teaching focuses on urban planning/governance, urban infrastructure and services, urban informalities, self-governance/urban politics, cities/towns.

Index

G. Alem Gebregiorgis et al. (eds.), *Planning Cities in Africa*, The Urban Book Series,
https://doi.org/10.1007/978-3-031-06550-7

N

GPSR Compliance

The European Union's (EU) General Product Safety Regulation (GPSR) is a set of rules that requires consumer products to be safe and our obligations to ensure this.

If you have any concerns about our products, you can contact us on ProductSafety@springernature.com

In case Publisher is established outside the EU, the EU authorized representative is:

Springer Nature Customer Service Center GmbH
Europaplatz 3
69115 Heidelberg, Germany

Batch number: 10852837

Printed by Printforce, the Netherlands